Edge Computing Systems with Kubernetes

A use case guide for building edge systems using K3s, k3OS, and open source cloud native technologies

Sergio Méndez

BIRMINGHAM—MUMBAI

Edge Computing Systems with Kubernetes

Group Product Manager: Rahul Nair
Publishing Product Manager: Preet Ahuja
Content Development Editor: Nihar Kapadia
Technical Editor: Shruthi Shetty
Copy Editor: Safis Editing
Project Coordinator: Ashwin Dinesh Kharwa
Proofreader: Safis Editing
Indexer: Pratik Shirodkar
Production Designer: Prashant Ghare
Senior Marketing Coordinator: Nimisha Dua

First published: October 2022

Production reference: 1280922

Published by Packt Publishing Ltd.
Livery Place
35 Livery Street
Birmingham
B3 2PB, UK.

ISBN 978-1-80056-859-4

www.packt.com

To my mother, Chusita, and my father, Arnaldo, my friends in the cloud native ecosystem, my colleagues at Yalo, and my students at USAC University, who have motivated and supported me throughout the process of writing this book.

Also, I would like to thank the Packt editors, who worked with me to ensure high-quality content.

Contributors

About the author

Sergio Méndez is a systems engineer and professor of operating systems in Guatemala at USAC University. His work at the university is related to teaching and researching cloud native technologies. He has experience working on DevOps and MLOps, using open source technologies at work. He is involved with several open source communities, including CNCF communities, promoting students in the CNCF ecosystem, and he hosts a cloud native meetup in Guatemala. He has been a speaker at several conferences, such as OSCON, KubeCon, WTF is Cloud Native?, and Kubernetes Community Days. He is also a Linkerd Ambassador.

I'd like to thank the team at Packt for giving me the opportunity to write my first book about something that has been wholly enjoyable. Most thanks, however, go to my reviewers and friends, Tiffany Jachja and Santiago Torres, for supporting me by reviewing this book during my busy professional life.

About the reviewers

Santiago Torres-Arias is an assistant professor at Purdue University's School of Electrical and Computer Engineering department. His interests include binary analysis, cryptography, distributed systems, and security-oriented software engineering. His current research focuses on securing the software development life cycle, cloud security, and update systems. Santiago is a member of the Arch Linux security team and has contributed patches to F/OSS projects at various degrees of scale, including Git, the Linux kernel, Reproducible Builds, NeoMutt, and the Briar project. Santiago is also a maintainer of the Cloud Native Computing Foundation's project **The Update Framework (TUF)**, as well as the lead of the in-toto project.

I'd like to thank the broader CNCF community for encouraging engagement from various perspectives and walks of life. In particular, I'd like to thank the leads of TAG-Security, as well as the Supply Chain Security Workgroup for all their input and feedback throughout the years. Outside of CNCF, I'd like to thank my colleagues and students at Purdue University for fostering a welcoming and truth-seeking environment.

Tiffany Jachja is an accomplished writer, speaker, and technologist, helping teams and other technologists deliver their best work. She brings her experiences in DevOps and cloud native application development to the data science field as an engineering leader. In her tenure within technology, she's led the successful delivery of technologies across various spaces and industries, including academia, government, finance, enterprise, start-ups, and media. She now helps people worldwide deliver their best work to create the success, recognition, and wealth they desire.

Table of Contents

3

K3s Advanced Configurations and Management   47

4

k3OS Installation and Configurations   65

5

K3s Homelab for Edge Computing Experiments   91

Part 2: Cloud Native Applications at the Edge

6

Exposing Your Applications Using Ingress Controllers and Certificates

7

GitOps with Flux for Edge Applications

8

Observability and Traffic Splitting Using Linkerd   167

9

Edge Serverless and Event-Driven Architectures with Knative and
Cloud Events   189

10

SQL and NoSQL Databases at the Edge   219

Part 3: Edge Computing Use Cases in Practice

11

Monitoring the Edge with Prometheus and Grafana   247

12

Communicating with Edge Devices across Long Distances Using LoRa   285

13

Geolocalization Applications Using GPS, NoSQL, and K3s Clusters   321

14

Computer Vision with Python and K3s Clusters   359

15

Designing Your Own Edge Computing System   401

Preface

Edge computing consists of processing data near to the source where this data is generated. In order to build an edge computing system, you must understand the different layers and components that an edge system uses to process the information. Using K3s a lightweight Kubernetes, you can take advantage of the use of containers to design distributed system and automate the way that your applications are updated. This book will gives you all the necessary tools to create your own edge system across learning the basics and different use cases of edge computing. By the end of this book, you will understand how to implement your own edge computing system that uses containers with K3s for your Kubernetes clusters and cloud native open source software.

Who this book is for

This book is for operations or DevOps engineers looking to move their data processing tasks to the edge or for those engineers looking to implement an edge computing system, but they don't have the technology background to do so. It can also be used for enthusiast and entrepreneurs looking to implement or experiment with edge computing for different or potential use case scenarios.

What this book covers

Chapter 1, Edge Computing with Kubernetes, explains basic concepts of Edge Computing including its components, layers, example architectures to build these kind of systems, and showing how to use cross compiling techniques for Go, Rust, Python and Java to run software at the edge that runs on devices with ARM processors.

Chapter 2, K3s Installation and Configuration, describes what K3s is, its components, and how to install K3s using different configurations such as single and multi-node, and finally explains advanced configurations for K3s clusters to use external storages to replace the use of etcd instead, expose applications outside the cluster installing and using an ingress controller, uninstalling the cluster and some useful commands to troubleshoot cluster installations.

Chapter 3, K3s Advanced Configurations and Management, introduce the reader to advanced configurations for its K3s cluster, including the installation of MetalLB a bare metal load balancer, the installation of Longhorn for storage at the edge, upgrades in the cluster and finally backing up and restoring K3s cluster configurations.

Chapter 4, k3OS Installation and Configurations, focuses on how to use k3OS a Kubernetes distribution packaged in an ISO image that could be used to be installed on edge devices. It also covers how to use overlay on ARM devices and perform installations using config files to configure a single or multi-node K3s clusters.

Chapter 5, K3s Homelab for Edge Computing Experiments, describes how to configure your own Homelab using all the previous configurations described in the previous chapters to produce a basic production ready environment to run your edge computing applications. Starting with cluster configurations, including configurations for ingress controller, persistence for applications and how to deploy a Kubernetes dashboard for your cluster at the edge.

Chapter 6, Exposing Your Applications Using Ingress Controllers and Certificates, gives an introduction about how to configure and use the ingress controllers NGINX, Traefik and Contour together with cert-manager to expose applications running on bare metal using TLS certificates.

Chapter 7, GitOps with Flux for Edge Applications, explores how to automate edge applications updates when source code changes are detected using a GitOps strategy together with Flux and GitHub Actions.

Chapter 8, Observability and Traffic Splitting Using Linkerd, describes how to use a Service Mesh to implement simple monitoring, observability, traffic splitting, and faulty traffic to improve services availability using Linkerd running at the edge.

Chapter 9, Edge Serverless and Event-Driven Architectures with Knative and Cloud Events, gives an introduction about how to implement your own serverless functions using Knative Serving. It also shows how to implement simple event-driven architectures using Knative Eventing together with Cloud Event to define and run events in your edge systems.

Chapter 10, SQL and NoSQL Databases at the Edge, explores different type of databases that can be used to record data at the edge. This chapter covers in specific the configuration and use of MySQL, Redis, MongoDB, PostgreSQL and Neo4j to cover different use cases for SQL and NoSQL databases running at the edge.

Chapter 11, Monitoring the Edge with Prometheus and Grafana, focuses on monitoring edge environments and devices using the time series database Prometheus and Grafana. In specific, this chapter focuses on creating custom real-time graphs for data coming from edge sensors that capture temperature and humidity.

Chapter 12, Communicating with Edge Devices across Long Distances Using LoRa, describes how to communicate edge devices in long distances using LoRa wireless protocol and how to visualize captured sensors edge data using MySQL and Grafana.

Chapter 13, Geolocalization Applications Using GPS, NoSQL, and K3s Clusters, describes how to implement a simple geolocalization or geo-tracking system using GPS modules and ARM devices showing vehicles moving in real-time, and reports of their tracking logs between a date range.

Chapter 14, Computer Vision with Python and K3s Clusters, describes how to create a smart traffic system to detect potential obstacles for drivers when driving in the city and give intelligent alerts and reports of the live state of traffic during rush hours. It is also described step by step how to implement this system using Redis, OpenCV, TensorFlow Lite, Scikit Learn and GPS modules running at the edge.

Chapter 15, Designing Your Own Edge Computing System, describes a basic methodology to create your own edge computing system and how you can use cloud provider managed services, complementary hardware and software and some useful recommendations while implementing your system. Finalizing with other use cases to explore for edge computing.

To get the most out of this book

To feel more comfortable with this book, you need some previous experience using Linux command line, and some basic programming knowledge. When reading a chapter, pay attention to download the source code, that will simplify the use of all the examples in this book.

This book mainly uses MacOS to perform local configurations. For the Raspberry Pi implementations Linux is used. Finally, there is a chapter that uses Windows to update the ESP32 firmware.

All the requirements need it to run the examples in this book are described in the Technical requirements section of each chapter.

If you are using the digital version of this book, we advise you to type the code yourself or access the code from the book's GitHub repository (a link is available in the next section). Doing so will help you avoid any potential errors related to the copying and pasting of code.

Download the example code files

You can download the example code files for this book from GitHub at `https://github.com/PacktPublishing/Edge-Computing-Systems-with-Kubernetes`. If there's an update to the code, it will be updated in the GitHub repository.

We also have other code bundles from our rich catalog of books and videos available at `https://github.com/PacktPublishing/`. Check them out!

Download the color images

We also provide a PDF file that has color images of the screenshots and diagrams used in this book. You can download it here: `https://packt.link/gZ68B`.

Conventions used

There are a number of text conventions used throughout this book.

`Code in text`: Indicates code words in text, functions, service name, deployment names, variables, pathnames, and URLs. Here is an example: "`WIFISetUp(void)`: we configure the Wi-Fi connection, here you have to replace `NET_NAME` with your network name and `PASSWORD` with the password to access your connection."

A block of code is set as follows:

```
@app.route('/')
def hello_world():
    return 'It works'
```

Any command-line input or output is written as follows:

```
$ mkdir code
$ kubectl apply -f example.yaml
```

Bold: Indicates a new term, an important word, or words that you see onscreen. For instance, words in menus or dialog boxes appear in **bold**. Here is an example: "Now create another file by clicking in **File | New**"

> **Tips or important notes**
> Appear like this.

Get in touch

Feedback from our readers is always welcome.

General feedback: If you have questions about any aspect of this book, email us at `customercare@packtpub.com` and mention the book title in the subject of your message.

Errata: Although we have taken every care to ensure the accuracy of our content, mistakes do happen. If you have found a mistake in this book, we would be grateful if you would report this to us. Please visit `www.packtpub.com/support/errata` and fill in the form.

Piracy: If you come across any illegal copies of our works in any form on the internet, we would be grateful if you would provide us with the location address or website name. Please contact us at `copyright@packt.com` with a link to the material.

If you are interested in becoming an author: If there is a topic that you have expertise in and you are interested in either writing or contributing to a book, please visit `authors.packtpub.com`.

Share Your Thoughts

Once you've read *Edge Computing Systems with Kubernetes*, we'd love to hear your thoughts! Scan the QR code below to go straight to the Amazon review page for this book and share your feedback.

https://packt.link/r/1-800-56859-2

Your review is important to us and the tech community and will help us make sure we're delivering excellent quality content.

Part 1:
Edge Computing Basics

In this part of the book, you will learn about the basic concepts, architectures, use cases, and current solutions for edge computing systems, as well as learning how to install a cluster using k3s/k3OS and Raspberry Pi devices.

This part of the book comprises the following chapters:

- *Chapter 1, Edge Computing with Kubernetes*
- *Chapter 2, K3s Installation and Configuration*
- *Chapter 3, K3s Advanced Configurations and Management*
- *Chapter 4, k3OS Installation and Configurations*
- *Chapter 5, K3s Homelab for Edge Computing Experiments*

1
Edge Computing with Kubernetes

Edge computing is an emerging paradigm of distributed systems where the units that compute information are close to the origin of that information. The benefit of this paradigm is that it helps your system to reduce network outages and reduces the delays when you process across the cloud. This means you get a better interactive experience with your machine learning or **Internet of Things (IoT)** applications. This chapter covers the basics and the importance of edge computing and how Kubernetes can be used for it. It also covers different scenarios and basic architectures using low-power devices, which can use private and public clouds to exchange data.

In this chapter, we're going to cover the following main topics:

- Edge data centers using K3s and basic edge computing concepts
- Basic edge computing architectures with K3s
- Adapting your software to run at the edge

Technical requirements

In this chapter, we are going to run our edge computing on an edge device (such as a **Raspberry Pi**), so we need to set up a cross-compiling toolchain for **Advanced RISC Machines (ARM)**.

For this, you need one of the following:

- A Mac with terminal access
- A PC with Ubuntu installed with terminal access
- A virtual machine with Ubuntu installed with terminal access

For more detail and code snippets, check out this resource on GitHub: `https://github.com/PacktPublishing/Edge-Computing-Systems-with-Kubernetes/tree/main/ch1`.

Edge data centers using K3s and basic edge computing concepts

With the evolution of the cloud, companies and organizations are starting to migrate their processing tasks to edge computing devices, with the goal to reduce costs and get more benefits for the infrastructure that they are paying for. As a part of the introductory content in this book, we must learn about the basic concepts related to edge computing and understand why we use K3s for edge computing. So, let's get started with the basic concepts.

The edge and edge computing

According to the Qualcomm and Cisco companies, the edge can be defined as *"anywhere where data is processed before it crosses the Wide Area Network (WAN)"*; this is the edge, but what is edge computing? A post by Eric Hamilton from Cloudwards.net defines edge computing as *"the processing and analyzing of data along a network edge, closest to the point of its collection, so that data becomes actionable."* In other words, edge computing refers to processing your data near to the source and distributing the computation in different places, using devices that are close to the source of data.

To add more context, let's explore the next figure:

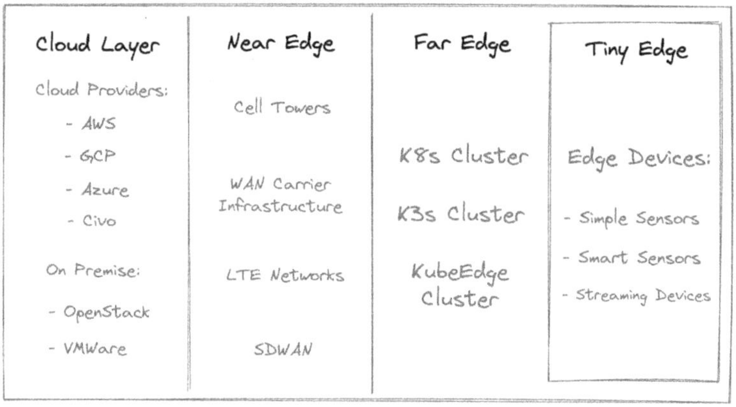

Cloud Layer	Near Edge	Far Edge	Tiny Edge
Cloud Providers: - AWS - GCP - Azure - Civo On Premise: - OpenStack - VMWare	Cell Towers WAN Carrier Infrastructure LTE Networks SDWAN	K8s Cluster K3s Cluster KubeEdge Cluster	Edge Devices: - Simple Sensors - Smart Sensors - Streaming Devices

Figure 1.1 – Components of edge layers

This figure shows how the data is processed in different contexts; these contexts are the following:

- **Cloud layer**: In this layer, you can find the cloud providers, such as AWS, Azure, GCP, and more.
- **Near edge**: In this layer, you can find telecommunications infrastructure and devices, such as 5G networks, radio virtual devices, and similar devices.

- **Far edge**: In this layer, you will find edge clusters, such as K3s clusters or devices that exchange data between the cloud and edge layer, but this layer can be subdivided into the tiny edge layer.

- **Tiny edge**: In this layer, you will find sensors, end-user devices that exchange data with a processing device, and edge clusters on the far edge.

> **Important Note**
> Remember that edge computing refers to data that is processed on edge devices before the result goes to its destination, which could be on a public or private cloud.

Other important concepts to consider for building edge clusters are the following:

- **Fog computing**: An architecture of cloud services that distribute the system across near edge and far edge devices; these devices can be geographically dispersed.

- **Multi-Access Edge Computing** (**MEC**): This distributes the computing at the edge of larger networks, with low latency and high bandwidth, and is the predecessor of mobile edge computing; in other words, the processing uses telecom networks and mobile devices.

- **Cloudlets**: This is a small-scale cloud data center, which could be used for resource-intensive use cases, such as data analytics, **Machine Learning** (**ML**) and so on.

Benefits of edge computing

With this short explanation, let's move on to understand the main benefits of edge computing; some of these include the following:

- **Reducing latency**: Edge computing can process heavy compute processes on edge devices, reducing the latency to bring this information.

- **Reducing bandwidth**: Edge computing can reduce the used bandwidth while taking part of the data on the edge devices, reducing the traffic on the network.

- **Reducing costs**: Reducing latency and bandwidth translates to the reduction of operational costs, which is one of the most important benefits of edge computing.

- **Improving security**: Edge computing uses data aggregation and data encryption algorithms to improve the security of data access.

Let's now discuss containers, Docker, and containerd.

Containers, Docker, and containerd for edge computing

In the last few years, container adoption has been increasing because of the success of Docker. Docker has been the most popular container engine for the last few years. Container technology gives businesses a way to design applications using microservices architecture. This way, companies speed up their development and strategies for scaling their applications. So, to begin with a basic concept: *A container is a small runtime environment that packages your application with all the dependencies needed for it to run.* This concept is not new, but Docker, a container engine, popularized this concept. In simple words, Docker uses small operating system images with the necessary dependencies to run your software. This can be called operating system virtualization. What this does is use the **cgroups** kernel feature of Linux to limit CPU, memory, network, I/O, and so on for your processes. Other operating systems, such as Windows or FreeBSD, use similar features to insulate and create this type of virtualization. Let's see the next figure to represent these concepts:

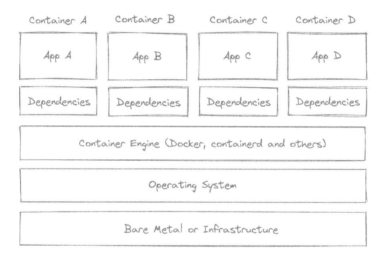

Figure 1.2 – Containerized applications inside the OS

This figure shows that a container doesn't depend on special features, such as a hypervisor that is commonly seen in hardware virtualization used by VMware, Hyper-V, and Xen; instead of that, the application runs as a binary inside the container and reuses the host kernel. Let's say that running a container is almost like running a binary program inside a directory but adds some resource limits, using cgroups in the case of Linux containers.

Docker implements all these abstractions. It is a popular container toolchain that adds some versioning features, such as Git. That was the main reason it became very popular, and it features easy portability and versioning at the operating system level. At the moment, containerd is the container runtime used by Docker and Kubernetes to create containers. In general, with containerd, you can create containers without extra features; it's very optimized. With the explosion of edge computing, containerd has become an important piece of software to run containers in low-resource environments.

In general, with all these technologies you can do the following:

- Standardize how to package your software.

- Bring portability to your software.

- Maintain your software in an easier way.

- Run applications in low-resource environments.

So, Docker must be taken into consideration as an important software piece to build edge computing and low-resource environments.

Distributed systems, edge computing, and Kubernetes

In the last decade, distributed systems evolved from multi-node clusters with applications using monolithic architectures to multi-node clusters with microservices architectures. One of the first options to start building microservices is to use containers, but once the system needs to scale, it is necessary to use an orchestrator. This is where Kubernetes comes into the game.

As an example, let's imagine an orchestra with lots of musicians. You can find musicians playing the piano, trumpets, and so on. But if the orchestra was disorganized, what would you need to have to organize all the musicians? The answer is an orchestra director or an orchestrator. Here is when Kubernetes appears; each musician is a container that needs to communicate or listen to other musicians and, of course, follow the instructions of the orchestra director or orchestrator. In this way, all the musicians can play their instruments at the right time and can sound beautiful.

This is what Kubernetes does; it is an orchestrator of containers, but at the same time it is a platform with all the necessary prebuilt pieces to build your own distributed system, ready to scale and designed with best practices that can help you to implement agile development and a DevOps culture. Depending on your use case, sometimes it's better to use something small such as Docker or containerd, but for complex or demanding scenarios, it's better to use Kubernetes.

Edge clusters using K3s – a lightweight Kubernetes

Now, the big question is how to start building edge computing systems. Let's get started with K3s. K3s is a Kubernetes-certified distribution created by Rancher Labs. K3s doesn't include by default extra features that are not vital to be used on Kubernetes, but they can be added later. K3s uses containerd as its container engine, which gives K3s the ability to run on low-resource environments using ARM devices. For example, you can also run K3s on x86_64 devices in production environments. However, for the purpose of this book, we will use K3s as our main piece of software to build edge computing systems using ARM devices.

Talking about clusters at the edge, K3s offers the same power as Kubernetes but in a small package and in an optimized way, plus some features designed especially for edge computing systems. K3s is very easy to use, compared with other Kubernetes distributions. It's a lightweight Kubernetes that can be used for edge computing, sandbox environments, or whatever you want, depending on the use case.

Edge devices using ARM processors and micro data centers

Now, it's time to talk about edge devices and ARM processors, so let's begin with edge devices. Edge devices are designed to process and analyze information near to the data source location; this is where the *edge* computing mindset comes from. Talking about low-energy consumption devices, x86 or Intel processors consume more energy and get warmer than ARM processors. This means more power and more cooling; in other words, you will pay more money for x86_64 processors. On the other hand, ARM processors have less computational power and consume less energy. That's the reason for the success of ARM processors on smartphone devices; they give you better cost and benefit between processing and energy consumption compared to Intel processors.

Because of that, companies are interested in designing micro data centers using ARM processors in their servers. For the same reason, companies are starting to migrate their workloads to be processed by devices using ARM processors. One example is the AWS Graviton2, which is a service that offers cloud instances using ARM processors.

Edge computing diagrams to build your system

Right now, with all the basic concepts of containers, orchestrators, and edge computing and its layers, we can focus on the five basic diagrams of edge computing configurations that you can use to design this kind of system. So, let's use K3s as our main platform for edge computing for the next diagrams.

Edge cluster and public cloud

This configuration shares and processes data between the public or private cloud with edge layers, but let's explain its different layers:

- **Cloud layer**: This layer is in the public cloud and its provider, such as AWS, Azure, or GCP. This provider can offer instances using Intel or ARM processors. For example, AWS offers the AWS Graviton2 instance if you need an ARM processor. As a complement, the public cloud can offer managed services to store data such as databases, storage, and so on. The private cloud could be in this layer too. You can find software such as VMware ESXi or OpenStack to provide this kind of service or instance locally. You can even choose a hybrid approach using the public and the private cloud. In general, this layer supports your far and tiny edge layers for storage or data processing.

- **Near edge**: In this layer, you can find network devices to move all the data between the cloud layer and the far layer. Typically, these include telco devices, 5G networks, and so on.

- **Far edge**: In this layer, you can find K3s clusters, similar lightweight clusters such as KubeEdge, and software such as Docker or containerd. In general, this is your local processing layer.

- **Tiny edge**: This is a layer inside the far edge, where you can find edge devices such as smartwatches, IoT devices, and so on, which send data to the far edge.

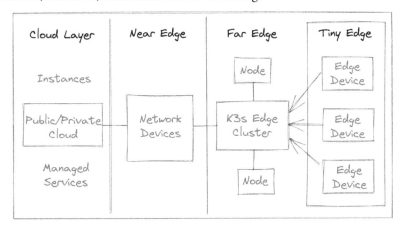

Figure 1.3 – Edge cluster and public cloud

Use cases include the following:

- Scenarios where you must share data between different systems across the internet or a private cloud

- Distribute data processing between your cloud and the edge, such as a machine learning model generation or predictions

- Scenarios where you must scale IoT applications, and the response time of the application is critical

- Scenarios where you want to secure your data using the aggregation strategy of distributing data and encryption across the system

Regional edge clusters and public cloud

This configuration is focused on distributing the processing strategy across different regions and sharing data across a public cloud. Let's explain the different layers:

- **Cloud layer**: This layer contains managed services such as databases to distribute the data across different regions.

- **Near edge**: In this layer, you can find network devices to move all the data between the cloud layer and the far layer. Typically, this includes telco devices, 5G networks, and so on.

- **Far edge**: In this layer, you can find K3s clusters across different regions. These clusters or nodes can share or update the data stored in a public cloud.

- **Tiny edge**: Here, you can find different edge devices close to each region where the far edge clusters process the information because of this distributed configuration.

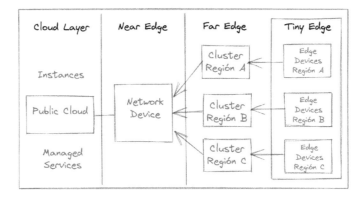

Figure 1.4 – Regional edge cluster and public cloud

Use cases include the following:

- Different cluster configurations across different regions
- Reducing application response time, choosing the closest data, or processing node location, which is critical in IoT applications
- Sharing data across different regions
- Distributing processing across different regions

Single node cluster and public/private cloud

This is a basic configuration where a single computer processes all the information captured on tiny edge devices. Let's explain the different layers:

- **Cloud layer**: In this layer, you can find the data storage for the system. It could be placed on the public or private cloud.
- **Near edge**: In this layer, you can find network devices to move all the data between the cloud layer and the far layer. Typically, this includes telco devices, 5G networks, and so on.
- **Far edge**: In this layer, you can find a single node K3s cluster that recollects data from tiny edge devices.
- **Tiny edge**: Devices that capture data, such as smartwatches, tablets, cameras, sensors, and so on. This kind of configuration is more for processing locally or on a small scale.

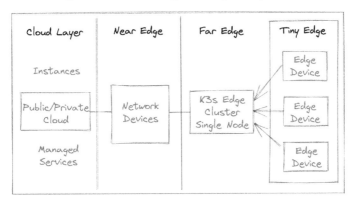

Figure 1.5 – Single node cluster and public/private cloud

Use cases include the following:

- Low-cost and low-energy consumption environments

- Green edge applications that can be powered by solar panels or wind turbines

- Small processes or use cases, such as analyzing health records or autonomous house systems that need something local or not too complicated

Let's now adapt the software to run at the edge.

Adapting your software to run at the edge

Something important while designing an edge computing system is to choose the processor architecture to build your software. One popular architecture because of the lower consumption for computing is ARM, but if ARM is the selected architecture, it is necessary to transform your current code in most of the cases from x86_64 (Intel) to ARM (ARMv7 such as RI and ARM such as AWS Graviton2 instances). The following subsections include short guides to perform the process to convert from one platform to another; this process is called cross-compiling. With this, you will be able to run your software on ARM devices using Go, Python, Rust, and Java. So, let's get started.

Adapting Go to run on ARM

First, it's necessary to install Go on your system. Here are a couple of ways to install Go.

Installing Go on Linux

To install Go on Linux, execute the following steps:

1. Download and untar the Go official binaries:

    ```
    $ wget https://golang.org/dl/go1.15.linux-amd64.tar.gz
    $ tar -C /usr/local -xzf go1.15.linux-amd64.tar.gz
    ```

2. Set the environment variables to run Go:

    ```
    $ mkdir $HOME/go
    ```

3. Set your GOPATH in the configuration file of your terminal with the following lines. ~/.profile is a common file to set these environment variables; let's modify the .profile file:

    ```
    $ export PATH=$PATH:/usr/local/go/bin
    $ export GOPATH=$HOME/go
    ```

4. Load the new configuration using the following command:

```
$ . ~/.profile
$ mkdir $GOPATH/src
```

5. (*Optional*). If you want to, you can set these environment variables temporarily in your terminal using the following commands:

```
$ export PATH=$PATH:/usr/local/go/bin
$ export GOPATH=$HOME/go
```

6. To check whether GOPATH is configured, run the following command:

```
$ go env GOPATH
```

Now, you are ready to use Go on Linux. Let's move to this installation using a Mac.

Installing Go on a Mac

To install Go on a Mac, execute the following steps:

1. Install Homebrew (called brew) with the following command:

```
$ /bin/bash -c "$(curl -fsSL https://raw.
githubusercontent.com/Homebrew/install/HEAD/install.sh)"
```

2. Once it is installed, install Go with brew:

```
$ brew install go
```

> **Important Note**
> To find out how to install brew, you can check the official page at https://brew.sh.

Cross-compiling from x86_64 to ARM with Go

To cross-compile from x86_64 to ARM, execute the following steps:

1. Create a folder to store your code:

```
$ cd ~/
$ mkdir goproject
$ cd goproject
```

2. Create an initial Go configuration to install external Go libraries outside the GOPATH command; for this, execute the next command:

```
$ go mod init main
```

3. Create the example.go file with Hello World as its contents:

```
$ cat << EOF > example.go
package main
import "fmt"
func main() {
    fmt.Println("Hello World")
}
EOF
```

4. Assuming that your environment is under x86_64 and you want to cross-compile for ARMv7 support, execute the following commands:

```
$ env GOOS=linux GOARM=7 GOARCH=arm go build example.go
```

Use the next line for ARMv8 64-bit support:

```
$ env GOOS=linux GOARCH=arm64 go build example.go
```

> **Important Note**
>
> If you want to see other options for cross-compiling, see https://github.com/golang/go/wiki/GoArm.

Set the execution permissions for the generated binary:

```
$ chmod 777 example
$ ./example
```

5. Copy the generated binary to your ARM device and test if it works.

In the next section, we will learn how to adapt Rust to run on ARM.

Adapting Rust to run on ARM

First, it's necessary to install Rust on your system. Here are a couple of ways to install Rust.

Installing Rust on Linux

To install Rust on Linux, execute the following steps:

1. Install Rust by executing the following command in the terminal:

    ```
    $ curl --proto '=https' --tlsv1.2 -sSf https://sh.rustup.
    rs | sh
    ```

2. Set the path for Rust in the configuration file of your terminal. For example, if you are using Bash, add the following line to your `.bashrc`:

    ```
    $ export PATH=$PATH:$HOME/.cargo/bin
    ```

Installing Rust on a Mac

To install Rust on a Mac, execute the following steps:

1. Install Homebrew with the following command:

    ```
    $ /bin/bash -c "$(curl -fsSL https://raw.
    githubusercontent.com/Homebrew/install/HEAD/install.sh)"
    ```

2. Once it is installed, install `rustup` with `brew`:

    ```
    $ brew install rustup-init
    ```

3. Run the `rustup` command to install Rust and all the necessary tools for Rust with the following command:

    ```
    $ rustup-init
    ```

4. Set your terminal environment variables by adding the following line to your terminal configuration file:

    ```
    $ export PATH=$PATH:$HOME/.cargo/bin
    ```

> **Important Note**
>
> Mac users often use the ZSH terminal, so they have to use `.zshrc`. If you are using another terminal, look for the proper configuration file or the generic `/etc/profile`.

Cross-compiling from x86_64 to ARMv7 with Rust on a Mac

To cross-compile from x86_64 to ARM, execute the following steps:

1. Install the complements to match the compiler and environment variables for ARMv7 architecture on your Mac; for this, execute the following command:

    ```
    $ brew tap messense/macos-cross-toolchains
    ```

2. Download the support for ARMv7 for cross-compiling by executing the following command:

    ```
    $ brew install armv7-unknown-linux-gnueabihf
    ```

3. Now set the environment variables:

    ```
    $ export CC_armv7_unknown_linux_gnueabihf=armv7-unknown-
    linux-gnueabihf-gcc
    $ export CXX_armv7_unknown_linux_gnueabihf=armv7-unknown-
    linux-gnueabihf-g++
    $ export AR_armv7_unknown_linux_gnueabihf=armv7-unknown-
    linux-gnueabihf-ar
    $ export CARGO_TARGET_ARMV7_UNKNOWN_LINUX_GNUEABIHF_
    LINKER=armv7-unknown-linux-gnueabihf-gcc
    ```

4. Create a folder to store your code:

    ```
    $ cd ~/
    $ mkdir rustproject
    $ cd rustproject
    ```

5. Create an initial Hello World project with Rust:

    ```
    $ cargo new hello-rust
    $ cd hello-rust
    ```

    The generated Rust code will look like this:

    ```
    fn main() {
      println!("Hello, world!");
    }
    ```

    The source code will be located at src/main.rs.

6. Add the support for ARMv7:

    ```
    $ rustup target add armv7-unknown-linux-gnueabi
    ```

7. Build your software:

```
$ cargo build --target=armv7-unknown-linux-gnueabi
```

8. Copy the binary file into your device and test whether it works:

```
$ cargo build --target=armv7-unknown-linux-gnueabi
```

9. The generated binary will be inside the `target/armv7-unknown-linux-gnueabi/hello-rust` folder.

10. Now copy your binary into your device and test whether it works.

> **Important Note**
>
> For more options for cross-compiling with Rust, check out `https://doc.rust-lang.org/nightly/rustc/platform-support.html` and `https://rust-lang.github.io/rustup/cross-compilation.html`. For the toolchain for Mac and AArch64 (64-bit ARMv8), check out `aarch64-unknown-linux-gnu` inside the repository at `https://github.com/messense/homebrew-macos-cross-toolchains`.

Adapting Python to run on ARM

First, it is necessary to install Python on your system. There are a couple of ways of doing this.

Installing Python on Linux

To install Python, execute the following steps:

1. Update your repositories:

```
$ sudo apt-get update
```

2. Install Python 3:

```
$ sudo apt-get install -y python3
```

Install Python on a Mac

To install Python on a Mac using Homebrew, execute the following steps:

1. Check for your desired Python version on brew's available version list:

```
$ brew search python
```

2. Let's say that you choose Python 3.8; you have to install it by executing the following command:

```
$ brew install python@3.8
```

3. Test your installation:

```
$ python3 --version
```

Cross-compiling from x86_64 to ARM with Python

Python is very important and one of the most popular languages now, and it is commonly used for AI and ML applications. Python is an interpreted language; it needs a runtime environment (such as Java) to run the code. In this case, you must install Python as the runtime environment. It has similar challenges running code as Java but has other challenges too. Sometimes, you need to compile libraries from scratch to use it. The standard Python libraries currently support ARM, but the issue is when you want something outside those standard libraries.

As a basic example, let's run Python code across different platforms by executing the following steps:

1. Create a basic file called example.py:

```
def main():
    print("hello world")
if __name__ == "__main__":
    main()
```

2. Copy example.py to your ARM device.

3. Install Python 3 on your ARM device by running the following command:

```
$ sudo apt-get install -y python3
```

4. Run your code:

```
$ python3 example.py
```

Adapting Java to run on ARM

When talking about Java to run on ARM devices, it is a little bit different. Java uses a hybrid compiler – in other words, a two-phase compiler. This means that it generates an intermediate code called bytecode and is interpreted by a **Java Virtual Machine (JVM)**. This bytecode is a cross-platform code and, following the Java philosophy of *compile once and run everywhere*, it means that you can compile using the platform you want, and it will run on any other platform without modifications. So, let's see how to perform cross-compiling for a basic Java program that can run on an ARMv7 and an ARMv8 64-bit device.

Installing Java JDK on Linux

To install Java on Linux, execute the following commands:

1. Update the current repositories of Ubuntu:

   ```
   $ sudo apt-get update
   ```

2. Install the official JDK 8:

   ```
   $ sudo apt-get install openjdk-8-jre
   ```

3. Test whether javac runs:

   ```
   $ javac
   ```

Installing Java JDK on a Mac

If you don't have Java installed on your Mac, follow the next steps:

1. (*Optional*) Download Java JDK from the following link and choose the architecture that you need, such as Linux, Mac, or Windows: https://www.oracle.com/java/technologies/javase-downloads.html.

2. (*Optional*) Download and run the installer.

   To test whether Java exists or whether it was installed correctly, run the following command:

   ```
   $ java -version
   ```

3. Test whether the compiler is installed by executing the following command:

   ```
   $ javac -v
   ```

Cross-compiling from x86_64 to ARM with Java

Java is a language that generates an intermediate code called bytecode, which runs on the JVM. Let's say that you have a basic code in a file called Example.java:

```
class Example {
    public static void main(String[] args) {
        System.out.println("Hello world!");
    }
}
```

To execute your code, follow these steps:

1. To compile it, use the following command:

    ```
    $ javac Example.java
    ```

    This will generate the intermediate code in a file called `Example.class`, which can be executed by the JVM. Let's do this in the next step.

2. To run the bytecode, execute the following command:

    ```
    $ java Example
    ```

3. Now, copy `Example.class` to another device and run it with the proper JVM using the `java` command.

Summary

This chapter explained all the basic concepts about edge computing and how it relates to other concepts, such as fog computing, MEC, and cloudlets. It also explained how containers and orchestrators such as Docker, containerd, and Kubernetes can help you to build your own edge computing system, using different configurations, depending on your own use case. At the end of the chapter, we covered how you can run and compile your software on edge devices using ARM processors, using the cross-compiling technique with Go, Python, Rust, and Java languages.

Questions

Here are a few questions to test your new knowledge:

1. What is the difference between the edge and edge computing?
2. What infrastructure configurations can you use to build an edge computing system?
3. How can containers and orchestrators help you to build edge computing systems?
4. What is cross-compiling and how can you use it to run your software on ARM devices?

Further reading

Here are some additional resources that you can check out to learn more about edge computing:

- *Near, Far or Tiny: Defining and Managing Edge Computing in a Cloud Native World, Keith Basil*: `https://vmblog.com/archive/2021/04/27/near-far-or-tiny-defining-and-managing-edge-computing-in-a-cloud-native-world.aspx`

- *What is Edge Computing: The Network Edge Explained, Eric Hamilton:, Cloudwards* (2018): `https://www.cloudwards.net/what-is-edge-computing`

- *IoT and Edge Computing for Architects – Second Edition, Perry Lea, Packt Publishing* (2020)

- *The IoT blog of Cisco*: `https://blogs.cisco.com/internet-of-things`

- *A secure data aggregation protocol for fog computing based smart grids*: `https://www.researchgate.net/publication/325638338_A_secure_data_aggregation_protocol_for_fog_computing_based_smart_grids.ng`

- *HarmonyCloud promotes edge computing implementation*: `https://www.cncf.io/blog/2021/08/31/harmonycloud-promotes-edge-computing-implementation`

- *Kubernetes – Bridging the Gap between 5G and Intelligent Edge Computing*: `https://www.cncf.io/blog/2021/03/01/kubernetes-bridging-the-gap-between-5g-and-intelligent-edge-computing`

- *CNCF YouTube video list of Kubernetes on Edge Day 2021*: `https://www.youtube.com/watch?v=W1v2Gb6URsk&list=PLj6h78yzYM2PuR1pP14DBLW7aku1Ia520`

- *Cross-Compiling using Rust for Mac*: `https://github.com/messense/homebrew-macos-cross-toolchains`

- *Cross-Compiling with Python*: `https://crossenv.readthedocs.io/en/latest/quickstart.html`

- *For instructions to download and install OpenJDK*: `https://openjdk.java.net/install`

2
K3s Installation and Configuration

This chapter offers a quick deep dive into K3s. We will start by understanding what K3s is and its architecture, and then we will learn how to prepare your ARM device for K3s. Following this, you will learn how to perform a basic installation of K3s from a single node cluster to a multi-node cluster, followed by a backend configuration using MySQL. Additionally, this chapter covers how to install an Ingress controller, using Helm Charts and Helm, to expose your Services across the load balancer created by NGINX. Finally, we will look at how to uninstall K3s and troubleshoot your cluster. At the end of the chapter, you will find additional resources to implement additional customizations for K3s.

In this chapter, we're going to cover the following main topics:

- Introducing K3s and its architecture
- Preparing your edge environment to run K3s
- Creating K3s single and multi-node clusters
- Using external MySQL storage for K3s
- Installing Helm to install software packages in Kubernetes
- Changing the default Ingress controller
- Uninstalling K3s from the master node or an agent node
- Troubleshooting a K3s cluster

Technical requirements

For this chapter, you will need one of the following options:

- Raspberry Pi 4 Model B with 4 GB of RAM (suggested minimum)

- An AWS account to create a **Graviton2** instance

- Any x86_64 VM instance with Linux installed

- An internet connection and DHCP support for local K3s clusters

With these requirements, we are going to install K3s and start experimenting with this Kubernetes distribution. So, let's get started.

Introducing K3s and its architecture

K3s is a lightweight Kubernetes distribution created by Rancher Labs. It includes all the necessary components inside a small binary file. Rancher removed all the unnecessary components for this Kubernetes distribution to run the cluster, and it also added other useful features to run K3s at the edge, such as MySQL support as a replacement for etcd, an optimized Ingress controller, storage for single node clusters, and more. Let's examine *Figure 2.1* to understand how K3s is designed and packaged:

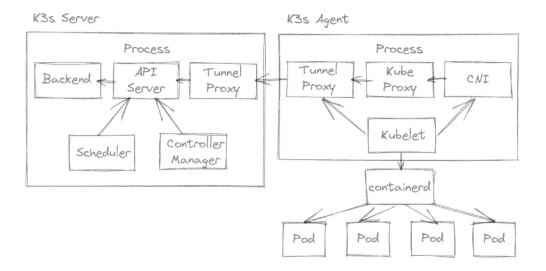

Figure 2.1 – The K3s cluster components

In the preceding diagram, you can see that K3s has two components: the server and the agent. Each of these components must be installed on a node. A node is a bare metal machine or a VM that works as a master or agent node. The master node manages and provisions Kubernetes objects such as Deployments, Services, and Ingress controllers inside the agent nodes. An agent node oversees the processing of information using these objects. Each node uses the different components shown in *Figure 2.1*, and they are provided in a single binary that packages all the necessary components to run the master and agent nodes. At the process level, the master node runs the K3s server, and the agent node runs the K3s agent. For each component, you will find a tunnel proxy to interconnect the master with the agent (that is, the worker nodes).

By default, the K3s *agent and master nodes run* Flannel as the default **Container Network Interface (CNI)** plugin. CNI is the specification for container networking, and the CNI plugins are the interface that is used to manage the network connectivity of containers. It also installs **containerd** as your container engine to create your Pods. One thing that the server and agent both share is that each component consists of a single binary around 100 MB that includes all minimal components to run each node. However, you can add additional components removed in K3s that are included in vanilla Kubernetes clusters, when you need them.

In terms of what the role of each node is, the master node is called the **control plane**, that is, the one that manages all the Kubernetes cluster configurations, networking, and more. In comparison, the agent node is called the **data plane** on which all the services, network traffic, and processing occur.

Preparing your edge environment to run K3s

Before installing K3s, you need to follow the next steps to configure a K3s master or agent for your ARM devices. So, let's get started.

Hardware that you can use

First, you must prepare your device. There are several options regarding how to do this. The first is to buy a Raspberry device to begin experimenting with to create a low-cost edge system. To buy this device, you need to take into consideration the following hardware specifications and components:

- The Raspberry Pi 4 Model B with at least 4 GB of RAM as an ARM device.
- A power supply of 5V and 3A is recommended.
- An Ethernet cable for the internet connection.
- A Micro HDMI to HDMI cable.
- A MicroSD card: SanDisk Extreme MicroSDHC UHS-1 A1 V30 32GB, or similar, is recommended.
- A MicroSD card reader.

This setup will give you the best bang for your buck. You might be thinking *why this configuration?* Well, let me quickly explain. The Raspberry Pi 4 Model B has a lot of improvements in terms of speed processing compared with previous versions. When talking about compatibility, the Raspberry Pi has an ARMv7 processor that is supported by many languages and programs. It also supports OSes for ARM64 or AArch64 processors that are used for devices with ARMv8 processors. This processors' architectures are supported in Raspberry B models. However, for more production-ready devices, you might want to look at an ARM 64-bit device, such as UDOO X86 II ULTRA, which has a 64-bit processor.

Moving on to the power supply, you need a device with 5 V and 3A to prevent slowing the Raspberry Pi down. You can use a 5 V/2.4 A, but a 5 V/3 A power supply works better for the Raspberry Pi 4 Model B. If you have the money, go for the 4 Model B with 8 GB of memory.

Finally, for the MicroSD card, select a high-speed card. This will perform better when you are running your software. SanDisk has a nice MicroSD card; just look at the read and write speed and use a MicroSD with at least 32 GB. And don't use Wi-Fi if possible; that's the reason behind using an Ethernet cable, so you can have a stable connection.

Linux distributions for ARM devices

There are several GNU/Linux distributions or OSes that you can use depending on your use case:

- **Raspbian**: This is the first distribution that you can use that is optimized for Raspberry devices. It is reliable and ready to use.

- **Ubuntu**: This distribution can be used on Raspberry devices or other ARM 64-bit devices, including x86_64 devices. One of the advantages of Ubuntu is that it can be found in all the major cloud providers such as AWS, Azure, and GCP.

- **Alpine**: This is a small distribution with minimal software, which is designed to be a tiny distribution. It can be used as your next project to customize your own distribution according to your project needs.

- **k3OS**: This is a tiny distribution designed to only run K3s on edge devices, but it's versatile.

There are other distributions, but you can use these as a quick start for your edge projects.

Installing Ubuntu inside your MicroSD card

Now it's time to install your OS. To install your Linux distribution inside your MicroSD, first, you must download Raspberry Pi Imager for your system. In this case, we are going to use the Mac version. You can download it at `https://www.raspberrypi.org/software`.

To begin installing the OS inside your Raspberry device, perform the following steps:

1. Install the binary from the previous link and open it; you should see something like this:

Figure 2.2 – The Raspberry Pi Imager menu

2. Click on the **CHOOSE OS** button to choose the Ubuntu Server 20.04 64-bit OS for ARM64, which can be found by navigating to the **Other general purpose OS | Ubuntu** menu:

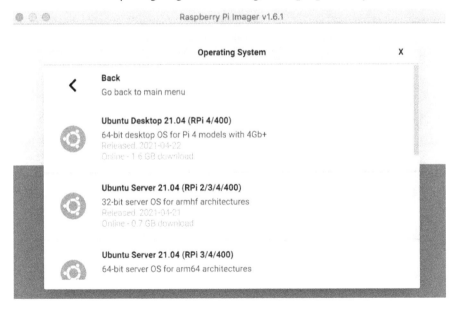

Figure 2.3 – The Raspberry distribution selection

3.   Next, insert your MicroSD card (you must buy an adapter to read MicroSD cards). Your device will appear when you select the **CHOOSE STORAGE** button:

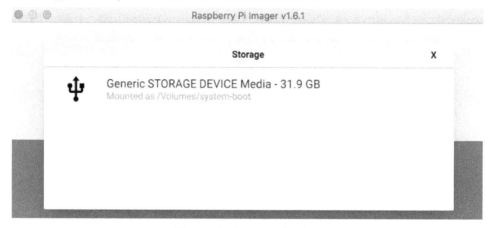

Figure 2.4 – Storage selection

4.   Then, click on the **WRITE** button:

Figure 2.5 – The last step to install the distribution onto your storage device

5.   Accept the option to write the device. Raspberry Pi Imager will then ask you for your username and password to continue writing to the MicroSD card:

Figure 2.6 – Confirmation to write to your MicroSD card

6.   Wait until the writing process finishes:

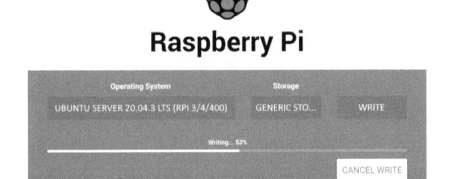

Figure 2.7 – Writing the OS onto the MicroSD card

7. Wait until the verifying process finishes:

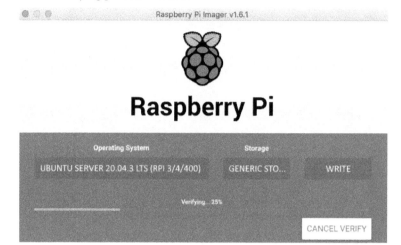

Figure 2.8 – Verifying that the OS has been written correctly

8. Extract your MicroSD card:

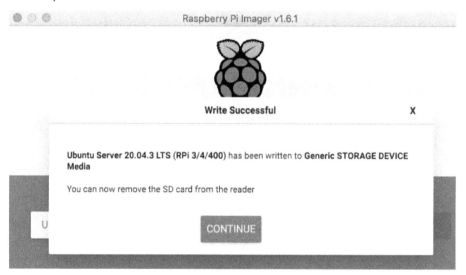

Figure 2.9 – Dialog showing when the writing process is complete

Now your MicroSD contains a fresh Ubuntu installation. In the next section, we will install K3s using this fresh installation.

Setting up Ubuntu before installing a K3s master or worker node

Right now, your device is prepared to run for the first time. Perform the following steps to configure and install it as a single node cluster:

1. Turn on your device.

2. When Ubuntu asks you for a username and password, enter the username and password as `ubuntu`; this is the default password for the first login.

3. Now, Ubuntu will ask you to change the default password. Let's use `k3s123` - as our password. Remember that in a real production scenario, you must use a stronger password.

4. Now, let's configure the network. By default, Ubuntu uses `init` cloud to configure the network. Let's deactivate this by creating a `99-disable-network-config.cfg` file with the following commands and content:

    ```
    $ sudo nano /etc/cloud/cloud.cfg.d/99-disable-network-
    config.cfg
    ```

    Here is the content of the file:

    ```
    network: {config: disabled}
    ```

5. If you execute `ifconfig`, you will see that your device is `eth0`. However, it could be named `es3` or something similar. So, let's modify the `50-cloud-init` file with the following command:

    ```
    $ sudo nano /etc/netplan/50-cloud-init.yaml
    ```

6. Next, modify the content of the file. It should look something like this:

    ```
    network:
      version: 2
      renderer: networkd
      ethernets:
        eth0:
          dhcp4: no
          addresses:
            - 192.168.0.11/24
          gateway4: 192.168.0.1
          nameservers:
              addresses: [8.8.8.8, 1.1.1.1]
    ```

> **Note**
>
> Remember that you should modify this file, as needed, by changing the address, gateway, and nameserver according to your current network or internet connection. For this local setup, we are using an internet connection with DHCP support.

7. Now apply the configuration, and you can reboot your device to determine whether your IP address is set when the OS starts. To do this, execute the following command:

    ```
    $ sudo netplan apply
    ```

8. Now configure the kernel parameters for the boot by editing the /boot/firmware/cmdline.txt file with the following command and content:

    ```
    $ sudo nano /boot/firmware/cmdline.txt
    ```

9. Add this content to the end of the line to enable container creation with containerd in your K3s cluster:

    ```
    cgroup_memory=1 cgroup_enable=memory
    ```

> **Note**
>
> If you are using Raspbian, this file is in /boot/cmdline.txt.

10. Edit the /etc/hostname file with a unique name, for example, master for your master node or worker-1, worker-2, and so on for the worker name using nano:

    ```
    $ sudo nano /etc/hostname
    ```

    Here is the content of the file:

    ```
    master
    ```

11. Edit the /etc/hosts file by adding the hostname. At the very least, you should have a line like this:

    ```
    $ sudo nano /etc/hosts
    ```

    The content, for example, could be as follows:

    ```
    127.0.0.1 localhost master
    ```

12. Now reboot your device:

    ```
    $ sudo reboot
    ```

This configuration is required to prepare your device to configure a K3s master node or agent node. In the next section, you will learn how to install K3s on your device.

Creating K3s single and multi-node clusters

In this section, you are going to learn how to configure K3s master and agent nodes on your Ubuntu OS for your ARM devices. To visualize what we are doing, let's take a closer look at *Figure 2.10*:

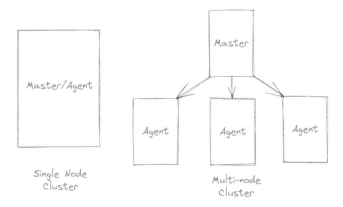

Figure 2.10 – The K3s cluster configurations

The preceding diagram shows that you can install a K3s cluster in the following configurations:

- **Single node cluster**: In this configuration, you only have one node that assumes the role of a master and agent/worker node at the same time. You can use this type of cluster for small applications. This is not ideal for heavy workloads, as it can slow down all the components. Remember that this node works as a master and an agent at the same time.

- **Multi-node cluster**: In this configuration, you have a master node that controls the agent/worker nodes; this configuration will be useful for high availability and heavy processing tasks.

With these brief descriptions, you can visualize what kind of configuration is required to create a K3s cluster. In the next section, you will learn how to create a single node cluster.

Creating a single node K3s cluster using Ubuntu OS

To begin installing K3s, you should use Ubuntu as your main distribution for K3s. You might be asking yourself why Ubuntu? Well, Ubuntu has a lot of pre-built features that can save some time when preparing your device. Additionally, it supports 32-bit and 64-bit ARM devices. I can recommend this distribution because of its compatibility and supported software. So, let's get started with this single node K3s cluster.

To install K3s (for a master-node or a single node cluster), you must perform the following steps:

1. Turn on your device and log in.

2. Once you are logged in, execute the following line in your Terminal to perform a basic installation of K3s:

```
$ curl -sfL https://get.k3s.io | INSTALL_K3S_EXEC="--
write-kubeconfig-mode 644 --no-deploy traefik --disable
traefik" sh -s -
```

> **Note**
>
> This command installs K3s without `traefik` as the default Ingress controller and gives you the ability to execute the `kubectl` command without using `sudo`. You can add some specific flags to use a specific version of K3s; please refer to the official documentation to learn more about this parameter. You can find the link at the end of this chapter.

3. (*Optional*) If you want to install K3s on AWS Graviton 2 instances or another cloud provider where the public IP is not associated with a network interface in the OS, you have to set the external IP parameter with the public IP of the instance, using the following commands:

```
$ PUBLIC_IP=YOUR_PUBLIC_IP|YOUR_PRIVATE_IP
$ curl -sfL https://get.k3s.io | INSTALL_K3S_EXEC="--
write-kubeconfig-mode 644 --no-deploy traefik --disable
traefik --tls-san "$PUBLIC_IP" --node-external-ip
"$PUBLIC_IP"" sh -s -
```

4. (*Optional*) If you want to implement a simple test, execute the following commands to expose a deployment using the `LoadBalancer` feature of K3s:

```
$ kubectl run nginx --image=nginx --restart=Never
$ kubectl expose pod/nginx --port=8001 --target-port=80
--type=LoadBalancer
```

Next, access the deployed `nginx` service using the public or private IP address of your K3s node on port `8001`; you can test the access by executing the following command:

```
$ curl http://YOUR_PUBLIC_OR_PRIVATE_IP:8001
```

Alternatively, if you have a private IP, run the following command:

```
$ curl http://YOU_PRIVATE_IP:8001
```

> **Note**
> This node will be a master node and an agent node at the same time.

Now we have installed a single node cluster. Let's go ahead and add more nodes to your new cluster in the next section.

Adding more nodes to your K3s cluster for multi-node configuration

So, what if you want to add more nodes to your single node cluster? To add more nodes to your cluster, first, you must follow the *Installing Ubuntu inside your MicroSD card* section for each new node. Then, you can continue with the following steps:

1. Log in to your master node:

    ```
    $ ssh ubuntu@MASTER_PUBLIC_OR_PRIVATE_IP
    ```

2. Extract the token to join the cluster from your master node using the following command:

    ```
    $ sudo cat /var/lib/rancher/k3s/server/node-token
    ```

3. Log out from your master node. Now you have the token to join additional nodes to the cluster.

    For each worker node to join the cluster, perform the following steps (this is the easier way).

4. Log in to your worker node that you want to add to the cluster:

    ```
    $ ssh ubuntu@WORKER_PUBLIC_OR_PRIVATE_IP
    ```

5. Set an environment variable with the token that your master generated:

    ```
    $ export TOKEN=YOUR_MASTER_TOKEN
    ```

6. Register your node using the following command:

    ```
    $ curl -sfL https://get.k3s.io | sh -s - agent --server
    https://MASTER_PUBLIC_OR_PRIVATE_IP:6443 --token ${TOKEN}
    --with-node-id
    ```

> **Note**
> If you have the same hostname for all your nodes, add the `--with-node-id` option and K3s will add a random ID at the end of your hostname so that you have a unique name for the nodes inside your cluster.

7. Exit from your worker node:

    ```
    $ exit
    ```

8. Log in to the master node:

    ```
    $ ssh ubuntu@MASTER_PUBLIC_OR_PRIVATE_IP
    ```

9. Check that your new node is running using the following command:

    ```
    $ kubectl get nodes
    ```

> **Note**
>
> You will have to wait a few minutes while the nodes change to the `Ready` state.

10. (*Optional*) If you have a different GNU/Linux distribution than Ubuntu, the following steps will work better with tiny distributions such as Alpine Linux. Log in to the worker node that you want to add to the cluster:

    ```
    $ ssh ubuntu@WORKER_PUBLIC_OR_PRIVATE_IP
    ```

11. Download the binary of K3s inside your worker node using the following command:

    ```
    $ curl -sfL https://github.com/k3s-io/k3s/releases/
    download/v1.21.1%2Bk3s1/k3s-arm64 > k3s > k3s | chmod +x
    k3s;sudo mv k3s /sbin
    ```

> **Note**
>
> Please navigate to `https://github.com/k3s-io/k3s/releases` to download the binary. Choose any method you wish to place this binary inside your worker node. The goal is to download the K3s binary inside your worker node. Note that in the previous command, version `v1.21.2+k3s1` was selected. So, modify the URL to fit your desired version.

12. Set an environment variable with the token that your master generated:

    ```
    $ export TOKEN=YOUR_MASTER_TOKEN
    $ sudo k3s agent --server https://myserver:6443
      --token ${TOKEN} --with-node-id &
    ```

13. Exit from your worker node:

    ```
    $ exit
    ```

14. Log in to your master node:

    ```
    $ ssh ubuntu@MASTER_IP
    ```

    If you want to set the role of your node, execute the following steps.

15. (*Optional*) Set the role of your new worker node using the following command:

    ```
    $ kubectl label nodes node_name kubernetes.io/role=worker
    ```

16. Exit from the master node:

    ```
    $ exit
    ```

Now you have a multi-node K3s cluster, and it's ready to use. In the next section, you will learn how to manage your cluster using the kubectl command.

Extracting K3s kubeconfig to access your cluster

Now, it's time to configure access to your K3s cluster from your computer using the kubectl command. To configure the connection of your new K3s cluster from the outside, perform the following steps:

1. Install the kubectl command by running the following commands for Mac installation:

    ```
    $ curl -LO https://dl.k8s.io/release/v1.22.0/bin/darwin/
    amd64/kubectl
    $ chmod +x ./kubectl
    $ sudo mv ./kubectl /usr/local/bin/kubectl
    $ sudo chown root: /usr/local/bin/kubectl
    ```

    Alternatively, if you are using Linux, run the following commands:

    ```
    $ curl -LO "https://dl.k8s.io/release/v1.22.0/bin/linux/
    amd64/kubectl"
    $ sudo install -o root -g root -m 0755 kubectl /usr/
    local/bin/kubectl
    ```

2. From the master node, copy the content inside /etc/rancher/k3s/k3s.yaml to your local ~/.kube/ config file

3. Take the following part of the server value:

    ```
    server: https://127.0.0.1:6443
    ```

    And change it to the following:

    ```
    server: https://MASTER_IP:6443
    ```

4. Change the permissions of this file using the following command:

```
$ chmod 0400 ~/.kube/config
```

5. Next, test whether you can access the cluster using the following command:

```
$ kubectl get nodes
```

This command returns the list of cluster nodes and their states.

> **Note**
>
> Remember to install the kubectl command-line tool before you copy the Rancher kubeconfig file onto your computer. Remember that the content of the k3s.yaml file has to be stored inside ~/.kube/config and it requires the 0400 permission. To learn how to install the kubectl command, navigate to https://kubernetes.io/docs/tasks/tools/install-kubectl-macos.

Now you are ready to perform more advanced configurations to create a new K3s cluster. Let's move on to the next section to learn more about this.

Advanced configurations

Now it's time to explore more advanced configurations that you can use to configure your K3s cluster at the edge.

Using external MySQL storage for K3s

K3s supports MySQL and SQLite, instead of etcd, as a data storage for your K3s cluster information. You can install MySQL in another node, a cloud instance, or a managed service on the cloud such as AWS Aurora or Google Cloud SQL. For example, let's attempt it with a cloud instance using DigitalOcean. However, you can do it on any cloud that you wish. So, let's get started with the following steps:

1. Log in to your cloud instance:

```
$ ssh root@IP_DATASTORE
```

2. Install Docker with the following commands:

```
$ apt-get update
$ apt-get install docker.io -y
$ docker run -d --name mysql -e MYSQL_ROOT_
PASSWORD=k3s123- \
-e MYSQL_DATABASE="k8s" -e MYSQL_USER="k3sadm" \
-e MYSQL_PASSWORD="k3s456-" \
```

```
-p 3306:3306 \
-v /opt/mysql:/var/lib/mysql \
mysql:5.7
```

3. Log out using the following command:

    ```
    $ exit
    ```

4. In your master node, execute the following:

    ```
    $ curl -sfL https://get.k3s.io | K3S_DATASTORE_
    ENDPOINT="mysql://k3sadm:k3s456-@tcp(YOUR_CLOUD_INSTANCE_
    IP:3306)/k8s" INSTALL_K3S_EXEC="--write-kubeconfig-mode
    644 --no-deploy traefik --disable traefik" sh -s -
    ```

> **Note**
>
> This will use the MySQL installation from your cloud instance. You must substitute YOUR_
> CLOUD_INSTANCE_IP with the IP of your cloud instance.

5. Extract the token to join the cluster from your master node with the following command:

    ```
    $ sudo cat /var/lib/rancher/k3s/server/node-token
    ```

6. Log out from your master node:

    ```
    $ exit
    ```

    For each worker node, execute the next step.

7. Install the agent to register and prepare your worker node:

    ```
    curl -sfL https://get.k3s.io | K3S_TOKEN=MASTER_TOKEN sh
    -s - agent --server https://MASTER_IP:6443
    ```

> **Note**
>
> You can execute kubectl get nodes to check your worker node has been added and is
> in the Ready state.

Now, you are ready to use your cluster with an external datastore instead of etcd or SQLite. In this case, we have a hybrid solution using local instances and a public instance to store the K3s configuration using MySQL. Remember that you can use MariaDB or another MySQL managed service from your favorite cloud provider. You can add multiple nodes configured as master nodes to your cluster for high availability in the main components of your cluster such as the Kubernetes API.

Installing Helm to install software packages in Kubernetes

Helm is a package manager for Kubernetes. With Helm, you can install software onto your Kubernetes cluster using a package definition called Helm Charts. You can use a public Helm Chart repository or your own repository to install packages. To install Helm in Linux or Mac, perform the following steps:

1. To install Helm on Linux, run the following commands:

    ```
    $ curl -fsSL -o get_helm.sh https://raw.
    githubusercontent.com/helm/helm/master/scripts/get-helm-3
    $ chmod 700 get_helm.sh
    $ ./get_helm.sh
    ```

2. To install Helm on Mac, run the following command:

    ```
    $ brew install helm
    ```

3. To begin installing Helm Charts, you should add a chart repository to Helm by running the following command on Linux or Mac:

    ```
    $ helm repo add bitnami https://charts.bitnami.com/
    bitnami
    ```

Now, let's examine how to change the default ingress controller.

Changing the default ingress controller

To begin this section, let's define what Ingress is and then define an Ingress controller. Based on the official Kubernetes website, an Ingress is a Kubernetes component that exposes your HTTP or HTTPS routes that match your internal services inside the cluster. A Service is an abstract way that Kubernetes uses to expose your application as a network service. And an Ingress controller is a component that is responsible for fulfilling the Ingress; this includes a load balancer that might also configure an edge router or proxy. There are a lot of implementations of Ingress controllers based on different edge routers or proxies such as Traefik, Envoy, Nginx, and more. By default, K3s includes Traefik version 1.0, which includes minimal features in which to route your services without consuming many resources.

If you want to use a different Ingress controller instead of the default option (**Traefik**), install the master node using the following commands:

1. Install the master node with the following parameters:

    ```
    $ curl -sfL https://get.k3s.io | INSTALL_K3S_EXEC="--
    write-kubeconfig-mode 644 --no-deploy traefik --disable
    traefik" sh -s -
    ```

2. Then, create a namespace to install the `nginx` Ingress controller with the following command:

```
$ kubectl create ns nginx-ingress
```

3. Add the Helm Charts repository:

```
$ helm repo add ingress-nginx \
  https://kubernetes.github.io/ingress-nginx
```

4. Update your repositories to get the latest version:

```
$ helm repo update
```

5. Install your Ingress controller:

```
$ helm install nginx-ingress \
  ingress-nginx/ingress-nginx \
  -n nginx-ingress
```

(*Optional*) If you want to test whether the `nginx-ingress` controller is working, follow the upcoming steps.

6. Create a deployment using the `nginx` image:

```
$ kubectl create deployment nginx --image=nginx
```

7. Expose the deployment using `ClusterIP`:

```
$ kubectl expose deployment/nginx --port=8001 \
  --target-port=80 --type=ClusterIP --name=nginx-srv
```

8. Create the `my-ingress.yaml` file using the following command:

```
apiVersion: networking.k8s.io/v1
kind: Ingress
metadata:
  name: my-ingress
  annotations:
    nginx.ingress.kubernetes.io/rewrite-target: /
spec:
  rules:
  - http:
      paths:
      - path: /mypath
```

```
            pathType: Prefix
            backend:
              service:
                name: nginx-srv
                port:
                  number: 8001
```

9. Create the Ingress using the following command:

    ```
    $ kubectl create -f my-ingress.yaml
    ```

10. Now test whether it works with the following command:

    ```
    $ curl http://LB_IP/my-path
    ```

> **Note**
>
> You must replace the value of LB_IP with the IP address of the LoadBalancer service created by the NGINX Ingress controller installation. In this case, is the same IP address of your master node.

11. To check the IP of where nginx-ingress has been exposed, execute the following command:

    ```
    $ kubectl get services -n nginx-ingress
    ```

> **Note**
>
> Take into consideration that K3s has its own behavior when using Kubernetes Services. To read more about this, please refer to https://rancher.com/docs/k3s/latest/en/networking.

Now that you understand how to install an Ingress controller and how to use it, it's time to learn how to uninstall K3s from your nodes if necessary.

Uninstalling K3s from the master node or an agent node

If you want to uninstall K3s in your master or agent nodes, you must execute the uninstall scripts provided by K3s installation. So, let's get started.

Uninstalling K3s from the agent node

To uninstall K3s from an agent (that is, the worker nodes), execute the following steps:

1. Log in to your agent node:

    ```
    $ ssh ubuntu@AGENT_NODE_IP
    ```

2. Uninstall the agent daemon and remove all the containers created on this node:

    ```
    $ k3s-agent-uninstall.sh
    $ sudo rm -R /etc/rancher
    $ sudo rm -R /var/lib/rancher
    ```

3. Log out from the agent node:

    ```
    $ exit
    ```

Uninstalling K3s from the master node

To uninstall K3s from the master node, execute the following steps:

1. Log in to your agent node:

    ```
    $ ssh ubuntu@MASTER_NODE_IP
    ```

2. Uninstall the agent daemon and remove all the containers created on this node:

    ```
    $ k3s-uninstall.sh
    $ sudo rm -R /etc/rancher
    $ sudo rm -R /var/lib/rancher
    ```

3. Log out from the agent node:

    ```
    $ exit
    ```

So, you have learned how to uninstall K3s, which could be useful when you want to try a new configuration with your devices. Now, let's move on to learn how to troubleshoot your cluster in the next section.

Troubleshooting a K3s cluster

This section includes some basic troubleshooting commands that you can use to test your cluster. There are different options for troubleshooting:

1. Execute the following command if you want to see the state of your nodes and check whether Kubernetes is running:

    ```
    $ kubectl get nodes
    ```

2. Create a pod to check whether your cluster can schedule pods:

    ```
    $ kubectl run nginx --image=nginx --restart=Never
    ```

3. Create a Service to expose the previously created Pod and test whether the LoadBalancer service works:

    ```
    $ kubectl expose pod/nginx --port=8001 \
      --target-port=80 \
      --type=LoadBalancer
    ```

4. Execute the following command if you want to check that the services and ports are working to expose your Services, which can be either LoadBalancer or NodePort:

    ```
    $ kubectl get services
    ```

5. Execute the following command if you want to check the logs in real time on your system:

    ```
    $ journalctl -f
    ```

6. Execute the following command to check whether the k3s service is running in your master node. This command must be executed inside your agent node:

    ```
    $ systemctl status k3s
    ```

7. Execute the following command to check whether the k3s-agent service is running in your agent/worker node. This command must be executed inside your agent node:

    ```
    $ systemctl status k3s-agent
    ```

> **Note**
>
> For more details about the different options and configurations available for K3s, you can visit https://rancher.com/docs/k3s/latest/en.

Summary

This chapter covered the firsts steps toward creating and customizing your Kubernetes cluster using the edge distribution of K3s. It also covered advanced configurations such as how to configure an external datastore for K3s that can help you to configure more robust and highly available solutions for edge K3s clusters. At the end of the chapter, we covered some advanced configurations such as how to install different Ingress controllers, the use of the Helm Chart operator, and basic troubleshooting commands for your cluster. With this knowledge, we can now jump to the next chapter to understand the advantage of k3OS to install K3s quickly and easily.

Questions

Here are a few questions to validate what you have learned in this chapter:

- What software can I use to prepare my ARM devices to install K3s?
- How can I install a basic multi-node cluster using K3s over ARM devices?
- How can I install a different Ingress controller?
- How can I use Helm to install packages in my cluster?
- How can I troubleshoot my cluster?

Further reading

You can refer to the following references for more information on the topics covered in this chapter:

- Raspberry Imager software: `https://www.raspberrypi.org/software`
- Ubuntu network configuration: `https://linuxize.com/post/how-to-configure-static-ip-address-on-ubuntu-20-04/#configuring-static-ip-address-on-ubuntu-server`
- The official documentation of K3s: `https://rancher.com/docs/k3s/latest/en`
- Installation options for K3s: `https://rancher.com/docs/k3s/latest/en/installation/install-options`
- Networking for K3s: `https://rancher.com/docs/k3s/latest/en/networking`
- The Helm website: `https://helm.sh`
- The K3s Helm Chart operator: `https://rancher.com/docs/k3s/latest/en/helm`
- Helm Charts Hub to find software that you want to install: `https://artifacthub.io`
- The official Kubernetes documentation: `https://kubernetes.io/docs`

K3s Advanced Configurations and Management

This chapter covers more advanced configurations for your K3s clusters. By default, K3s includes a load balancer called KlipperLB, but it has some limitations. For example, you don't have to repeat a port while creating a service, and it affects the way that you use a regular load balancer and NodePort service. It works well for simple deployments. In case you need another load balancer instead of Klipper, we cover how to install **MetalLB**, a bare metal load balancer. Then, we cover how to use advanced storage configuration to support read/write access modes for storage volumes with Longhorn, substituting the default local storage class provided by K3s. After this, we will do some common cluster management, including upgrading K3s, backing up, and restoring the cluster.

In this chapter, we're going to cover the following main topics:

- Bare metal load balancer with MetalLB
- Setting up Longhorn for storage
- Upgrading your cluster
- Backing up and restoring your K3s configurations

Technical requirements

For this chapter, you need the following:

- Raspberry Pi 4 model B with 4 GB RAM (minimum suggested)
- A cloud server or VM with Ubuntu 20.04 LTS
- Helm v3 installed in your device or client

With this, we are ready to learn this advanced configuration for K3s. So, let's get started.

Bare metal load balancer with MetalLB

In this section, you are going to explore MetalLB as a bare metal load balancer, which can give you powerful features to expose your services at the edge.

Load balancer services in Kubernetes

Before starting with KlipperLB, it's necessary to give some context about load balancers in Kubernetes. Kubernetes uses services to communicate or access your application. A ClusterIP service creates a DNS record, so this service could be reachable from within the cluster. A NodePort service exposes the service on each node's IP at a static port. This port is in the range of 30000–32767. And, finally, Kubernetes supports a load balancer service that exposes the service externally using a cloud provider's load balancer. In the case of K3s, it's going to use KlipperLB by default.

KlipperLB and MetalLB as bare metal load balancers

Edge devices and edge computing don't have a lot of resources, so it is common to find clusters that only have a single node. Generally, a Kubernetes load balancer service depends on the implementation of a specific cloud provider. It also works in layer 4 (the transport layer) to transmit **Transmission Control Protocol** (**TCP**) and **User Datagram Protocol** (**UDP**) protocols, and this load balancer is connected to a NodePort service too. So, in the case of edge devices, K3s implements KlipperLB.

KlipperLB works really nicely on low-resource devices or environments as k3s' load balancer. But when you have multi-node clusters, maybe KlipperLB doesn't offer the best features for service availability. That's where MetalLB comes into the game. KlipperLB and MetalLB offer a bare metal load balancer service on Kubernetes. In this case, you can use these implementations on K3s.

KlipperLB and MetalLB – the goods and the bads

Now, let's mention the pros and cons of each of those bare metal load balancers in terms of the implementation, dependencies, and use case. So, let's get started with KlipperLB.

The pros of KlipperLB are as follows:

- Pretty lightweight
- Simple to use with enough features for single node clusters

The cons of KlipperLB are as follows:

- Depends on **hostPort** or available ports to expose a pod.

- If the port is not available, the load balancer service stays on the pending state.

Talking about MetalLB, it uses layer 2 (the data link layer) where the format of data is defined. In this way, MetalLB uses a node for load balancing and has its own advantages and disadvantages. The next table summarizes this information:

Advantages	Disadvantages
- Can reassign a new node for load balancing if the nodes fail - Doesn't depend on available ports in the node - Announcement using ARP or BGP	- The load balancer node could be a point of failure - It needs multiple IPs per node - Dynamic IP assignment - Additional and complex installation to activate BGP support

In general, choose KlipperLB if you have a single node cluster and you want to avoid complex installations that use unique ports. Use MetalLB for multi-node clusters or installations where you can reuse ports and a more robust load balancer service.

Installing MetalLB

You need a K3s installation with the `--disable servicelb` option; if you have a previous installation, you have to reinstall K3s. To install K3s with this option, follow these steps:

1. Log in to your **virtual machine** (**VM**) or device using the following command:

   ```
   $ ssh ubuntu@YOUR_VM_IP
   ```

2. Install K3s using the following line. This applies to a simple ARM device for a basic installation without installing KlipperLB:

   ```
   $ curl -sfL https://get.k3s.io | INSTALL_K3S_EXEC="--write-kubeconfig-mode 644 --no-deploy traefik --disable traefik --disable servicelb" sh -s -
   ```

   (*Optional*) Install K3s using the following lines. First, set the `PUBLIC_IP` environment variable with the IP of your device or VM:

   ```
   $ PUBLIC_IP=YOUR_PUBLIC_IP|YOUR_PRIVATE_IP
   ```

Then, use the next lines to install K3s in a node that has a public IP:

```
$ curl -sfL https://get.k3s.io | INSTALL_K3S_EXEC="--
write-kubeconfig-mode 644 --no-deploy traefik --disable
traefik --tls-san "$PUBLIC_IP" --node-external-ip
"$PUBLIC_IP" --disable servicelb" sh -s -
```

3. Create a MetalLB namespace (`metallb-system`) with the official manifests, executing the following lines:

```
$ kubectl apply -f https://raw.githubusercontent.com/
metallb/metallb/v0.10.2/manifests/namespace.yaml
```

4. Before running the command to install MetalLB, you have to create a ConfigMap called `metallb-config` inside the `metallb-system` namespace. Let's call this file `config.yaml` with the following content:

```
apiVersion: v1
kind: ConfigMap
metadata:
  namespace: metallb-system
  name: config
data:
  config: |
    address-pools:
    - name: default
      protocol: layer2
      addresses:
      - 192.168.0.240-192.168.0.250
```

5. Now, create the ConfigMap by executing the following command:

```
$ kubectl apply -f config.yaml
```

6. Install MetalLB with the official manifests by executing the following lines:

```
$ kubectl apply -f https://raw.githubusercontent.com/
metallb/metallb/v0.10.2/manifests/metallb.yaml
```

Now that MetalLB is installed using YAML files, let's continue with the installation using Helm instead of YAML files.

> **Important Note**
> If you want to delete this or other installations, use the `delete` option instead of `apply` using the same command – for example, `kubectl delete -f YOUR_YAML_FILE`.

In case you want to install MetalLB using Helm v3, follow these steps:

1. Add the Helm Chart repository of MetalLB using the following commands:

    ```
    $ helm repo add metallb https://metallb.github.io/metallb
    ```

2. Install MetalLB using Helm by executing the following command:

    ```
    $ helm install metallb -n metallb-system metallb/metallb
    ```

3. If you want to install MetalLB with the `values.yaml` file, execute the following lines:

    ```
    $ helm install metallb -n metallb-system metallb/metallb
    -f values.yaml
    ```

4. You have to create the `values.yaml` file, with the following example content:

    ```
    configInline
      address-pools:
        - name: default
          protocol: layer2
          addresses:
          - 192.168.0.240-192.168.0.250
    ```

5. Now, you have to create the ConfigMap based on the installation using `kubectl` and change the namespace to `metallb-system` and the name to `metallb-config`. Then, apply YAML:

    ```
    $ kubectl apply -f config.yaml
    ```

> **Important Note**
> The `addresses` field corresponds to the range of IP addresses that MetalLB will use to assign to your services every time that you create a `LoadBalancer` service in Kubernetes.

6. Now, MetalLB is installed and ready to use.

Now you have a fresh installation of MetalLB ready to use. Now you have to learn how to troubleshoot MetalLB in the next section.

Troubleshooting MetalLB

Sometimes, it's necessary to troubleshoot our installations. If you are having trouble with your installation, here are some commands that you can use to troubleshoot a new installation of MetalLB. The following are steps and commands that you can use for this:

1. Log in to your VM or device:

    ```
    $ ssh ubuntu@NODE_IP
    ```

2. Create a pod to check whether your cluster can schedule pods:

    ```
    $ kubectl run nginx --image=nginx --restart=Never
    ```

3. Create a service to expose the pod created previously and test whether the `LoadBalancer` service works:

    ```
    $ kubectl expose pod/nginx --port=8001 \
      --target-port=80 \
      --type=LoadBalancer
    ```

4. Execute the following command if you want to check whether the services and port work to expose your services, which can be either `LoadBalancer` or `NodePort`:

    ```
    $ kubectl get services
    ```

5. Now, perform an access check for the assigned external IP to the NGINX service and execute the following command to check that MetalLB exposed your service:

    ```
    $ curl http://EXTERNAL_IP:8001
    ```

In case you want to check the logs of MetalLB in case of errors, look at the next pods inside the `metallb-system` namespace:

*   Controller
*   Speaker

Now you know how to do basic troubleshooting of MetalLB. Let's move to a more advanced storage configuration using Longhorn in the next section.

Setting up Longhorn for storage

In terms of persistent information, you will find two types of containers, stateless and stateful containers. A stateless or ephemeral container doesn't persist information generated inside a container. A stateful container can persist the information even when this is deleted. K3s includes, by default, a way to persist data using a storage type (called **storage class** in Kubernetes) called **local-path**. This storage is a basic and pretty lightweight implementation, designed for edge devices. A common feature used on Kubernetes is to have a persistent volume claim that allows your pods to consume (write and read data) from different nodes. And this is a persistence volume configuration with the access mode key, set as **ReadWriteMany** (**RWX**). This feature is often used in production scenarios and it's pretty important because it enables you to share information from your different services. Longhorn provides this feature in a pretty lightweight presentation and it's optimized for edge devices. Let's move to learn what Longhorn is and how you can install it.

Why use Longhorn?

Longhorn is designed to be a distributed and hyper-converged storage stack. Hyper-converged storage means that virtualization software abstracts and pools storage. Longhorn doesn't use a lot of resources, which gives you the ability to use it for advanced storage in edge devices. You can even simplify your workflows of snapshots, backups, and even disaster recovery. So, if you are looking for lightweight and advanced edge solutions for storage, Longhorn can fit your needs. There are other options, such as Rook, but Longhorn is an easy piece of software that can give you extra storage power without having to sacrifice resource consumptions. So, let's move on to learn how to install it and create a simple **persistent volume claim** for a pod in the next section.

Installing Longhorn with ReadWriteMany mode

To install Longhorn, follow these steps:

1. Log in to your VM or device:

    ```
    $ ssh ubuntu@NODE_IP
    ```

2. If you want to install the ReadWriteMany PVC mode, you have to install `nfs-common` on each VM with Ubuntu installed in your cluster. For this, execute the following command:

    ```
    $ sudo apt install -y nfs-common
    ```

3. Apply the official Longhorn manifests, as follows:

    ```
    $ kubectl apply -f https://raw.githubusercontent.com/
    longhorn/longhorn/v1.3.1/deploy/longhorn.yaml
    ```

> **Important Note**
> Longhorn will be installed in the `longhorn-system` namespace.

4. Create a `pvc.yaml` file:

```
apiVersion: v1
kind: PersistentVolumeClaim
metadata:
  name: longhorn-volv-pvc
spec:
  accessModes:
    - ReadWriteMany
  storageClassName: longhorn
  resources:
    requests:
      storage: 2Gi
```

5. Apply the `pvc.yaml` file:

```
$ kubectl create -f pvc.yaml
```

> **Important Note**
> You can use different PVC modes such as `ReadWriteOnce` or `ReadOnlyMany`. By default, the storage classes at least support `ReadWriteOnce`. So, `ReadWriteMany` is a special feature that uses `NFS` and is included in Longhorn.

Now, it's time to create a pod using this PVC using the Longhorn storage class. To do this, follow these steps:

6. Create the `pod.yaml` file to create a pod using the previously created PVC:

```
echo "
apiVersion: v1
kind: Pod
metadata:
  name: volume-test
  namespace: default
spec:
  containers:
```

```
    - name: volume-test
      image: nginx:stable-alpine
      imagePullPolicy: IfNotPresent
      volumeMounts:
      - name: volv
        mountPath: /data
      ports:
      - containerPort: 80
    volumes:
    - name: volv
      persistentVolumeClaim:
        claimName: longhorn-volv-pvc" > pod.yaml
```

7. Apply the pod.yaml file to create the pod:

    ```
    $ kubectl create -f pod.yaml
    ```

Now, you have Longhorn installed and running. So, let's move on to learn how to use the Longhorn UI in the next section.

Using Longhorn UI

If you want to access the Longhorn UI, you have to check the services created on longhorn-system and execute a port-forward; if you installed MetalLB, you can create a LoadBalancer service to expose the Longhorn UI.

To access Longhorn with a port-forward, execute the following steps:

1. Run the next port-forward command locally in order to access the UI in your browser:

    ```
    $ kubectl port-forward svc/longhorn-frontend -n longhorn-
    system 8080:80
    ```

2. Now, open your browser at `http://localhost:8080`; you will see the following dashboard:

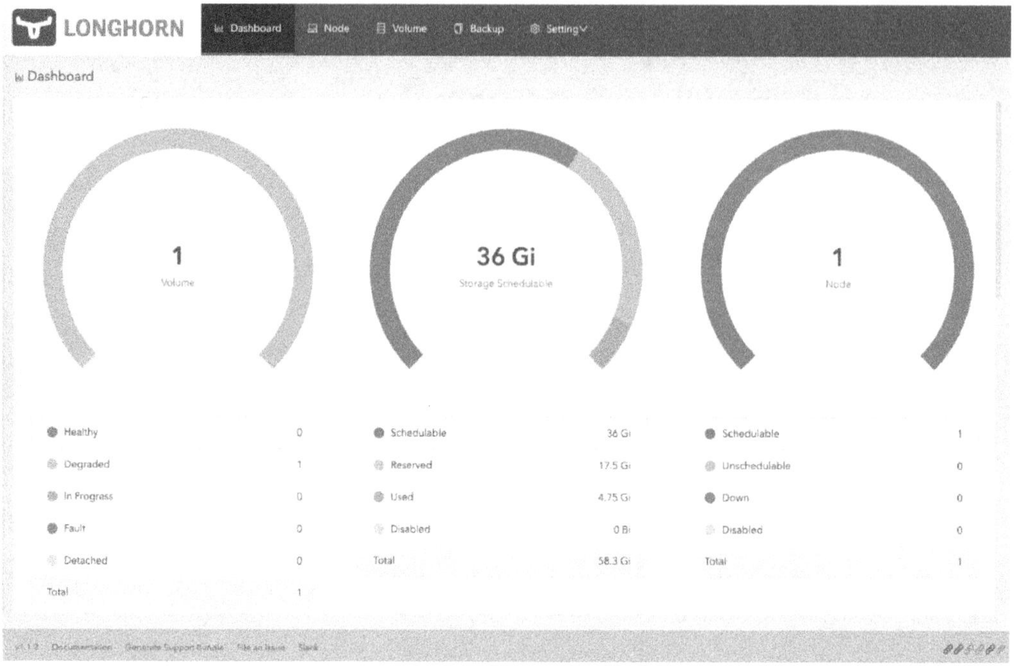

Figure 3.1 – Longhorn UI

With this dashboard, you can manage your **Persistent Volume Claims** (**PVCs**) using the UI; for more references, you can visit the following link: `https://longhorn.io/docs/1.3.1/deploy/accessing-the-ui`.

Now you know how to install and use Longhorn. Let's go ahead and do some basic troubleshooting.

Troubleshooting Longhorn

Using the preceding example as reference, to troubleshoot the PVC creation using Longhorn, you can use the following commands:

1. Check whether the Longhorn pods are running successfully with the following command:

    ```
    $ kubectl get pods -n longhonr-system
    ```

2. Check whether the PV was created:

    ```
    $ kubectl get pv
    ```

3. Check whether the PVC was created:

   ```
   $ kubectl get pvc
   ```

4. Check whether the pod from `pod.yaml` using the new Longhorn storage class was created:

   ```
   $ kubectl get pods
   ```

With these commands, you can find errors that come up when a pod or deployment uses a PVC with the Longhorn storage class.

> **Important Note**
>
> The previous four commands will return errors in case something goes wrong. For more information about this, you can check `https://kubernetes.io/docs/tasks/configure-pod-container/configure-persistent-volume-storage` or `https://kubernetes.io/docs/concepts/storage/persistent-volumes/#class`.

Now, we are ready to learn another advanced topic about upgrading the cluster. So, let's move to the next section.

Upgrading your cluster

Sometimes, you want to be up to date with the new versions and features of K3s. The next sections explain how to perform these upgrading processes.

Upgrading using K3s Bash scripts

To perform an upgrade in your nodes, you have to follow these steps:

1. First, you have to stop K3s on your device with the following command:

   ```
   $ /usr/local/bin/k3s-killall.sh
   ```

2. Now, you have to choose the version which you want to upgrade to. In general, there are three options – choose the latest or most stable channel, or pick a specific version. The next command will update your cluster to the latest stable version available:

   ```
   $ curl -sfL https://get.k3s.io | sh -
   ```

3. Now, if you want to update to the latest version, which is not so stable, you can execute the following command:

```
$ curl -sfL https://get.k3s.io | INSTALL_K3S_
CHANNEL=latest sh -
```

4. The last option is to pick a specific version. For this, you have to execute the following command:

```
$ curl -sfL https://get.k3s.io | INSTALL_K3S_
VERSION=vX.Y.Z-rc1 sh -
```

> **Important Note**
>
> You can visit `https://update.k3s.io/v1-release/channels` to check the latest, stable, or specific available version of K3s or the official site of k3s at `https://k3s.io` in the GitHub section.

Now you know how to upgrade your cluster using the K3s scripts. Let's move on to learn this manually in the next section.

Upgrading K3s manually

If you want to perform a manual upgrading of the K3s version, you can follow the following steps, based on the official K3s website documentation:

1. Download your desired version of the K3s binary from releases. To do this, check this link: `https://github.com/k3s-io/k3s/releases`.

2. Copy the downloaded binary to the `/usr/local/bin` folder.

3. Stop the old k3s binary. For this, you can execute the following command:

```
$ /usr/local/bin/k3s-killall.sh
```

4. Delete the old binary.

5. Launch the new K3s binary with the next command:

```
$ sudo systemctl restart k3s
```

Now, you know how to do the K3s manually, but there is something that you have to know, and that is to restart the service to apply the next changes. This is covered in the next section.

Restarting K3s

When you perform software or hardware upgrades, or when a restart is needed to fix errors, you can restart K3s services using `systemd` and **OpenRC**.

To restart K3s using `systemd`, follow these steps:

1. To restart the K3s service in your master node, execute the following command:

   ```
   $ sudo systemctl restart k3s
   ```

2. To restart the K3s agent service in your agent nodes, execute the following command:

   ```
   $ sudo systemctl restart k3s-agent
   ```

To restart K3s using OpenRC, follow these steps:

1. To restart the K3s service in your master node, execute the following command:

   ```
   $ sudo service k3s restart
   ```

2. To restart the K3s-agent service in your agent nodes, execute the following command:

   ```
   $ sudo service k3s-agent restart
   ```

Now that you know all the necessary steps to upgrade your K3s cluster, it's time to move on to other advanced topics – backups and restorations. Let's move on to the next section to learn about this.

Backing up and restoring your K3s configurations

Backups and restoration of your Kubernetes objects are something to consider in production environments. This section explains how to perform these kinds of tasks for the default storage, SQLite, how to install and manage **etcd** on K3s, and basic resources if you are using the SQL backends of K3s.

Backups from SQLite

If you are using the default storage, SQLite, follow these steps:

1. Log in to your master node:

   ```
   $ ssh ubuntu@NODE_IP
   ```

2. Stop the K3s using the following command:

   ```
   $ k3s-killall.sh
   ```

3. Change to the `/var/lib/rancher/k3s/` directory server:

```
$ sudo cd /var/lib/rancher/k3s
```

4. Copy the folder server inside the k3s folder:

```
$ sudo cp -R /var/lib/rancher/k3s folder_of_destination
```

5. Download this folder on another device if necessary.

Backups and restoring from the SQL database K3s backend

If you are using external storage – let's say, for example, MySQL – you have to use a tool or the command to back up your database.

Backing up MySQL

In the case of MySQL, you can execute the following steps to back up your K3s configurations:

1. Get your database credentials to use the mysqldump command.

2. Run the following command to back up your database, which in this case is called k3s, using the YOUR_USER user, the YOUR_PASSWORD password, and an output file called output. sql from the YOUR_HOST host:

```
$ mysqldump -h YOUR_HOST -u YOUR_USER -pYOUR_PASSWORD k3s
> output.sql
```

> **Important Note**
>
> You can modify the YOUR_HOST, YOUR_USER, and YOUR_PASSWORD values, the database name instead of k3s, and even the name of the output file to customize your backup command. The –h option can be optional if you are connected to the same host where the database is installed. By default, it connects to localhost. You can check this link for other examples to back up your MySQL: https://www.tecmint.com/mysql-backup-and-restore-commands-for-database-administration.

Now the backup is ready to be used. In the next section, you are going to use the backup to restore your database.

Restoring MySQL

Now it is time to restore your database. Follow the next steps for the restoration:

1. Get your database credentials to use the database with the `mysql` command.

2. Run the following command to restore your database backup. We are using the `k3s` database. Change YOUR_HOST and YOUR_PASSWORD parameters according to the database used as data storage for your `k3s` cluster. Finally, the `output.sql` file is used to load your backup and restore your database:

    ```
    $ mysql -h YOUR_HOST -u YOUR_USER -pYOUR_PASSWORD k3s <
    output.sql
    ```

> **Important Note**
> You can modify the values from the previous command to perform your restoration with the `output.sql` file.

Backing up and restoring other data storages

If you are using other K3s backends, such as PostgreSQL or `etcd`, you can check the official documentation for each database.

For PostgreSQL, check the following link: `https://www.postgresql.org/docs/8.3/backup-dump.html`.

For `etcd`, check the following link: `https://etcd.io`.

Now that you have learned how to restore your MySQL data storage for your K3s cluster, let's move on to the next section to understand how to use `etcd` as your data storage.

Embedded etcd management

`etcd` is the default type of storage to store all the Kubernetes objects in your cluster. `etcd`, by default, was removed from K3s, but you can install it. K3s customized how `etcd` works for your cluster; this includes some custom features that you can't find in a regular Kubernetes cluster that uses `etcd`. So, let's get started with installing `etcd` in K3s.

Installing the etcd backend

If you want to install it, follow these steps:

1. To install K3s with the `etcd` backend, you have to execute the following command to include `etcd` in the K3s installation. This has to be executed in the master node:

    ```
    $ curl -sfL https://get.k3s.io | INSTALL_K3S_EXEC="server
    --cluster-init" sh -s -
    ```

2. Set your TOKEN variable, with the YOUR_TOKEN master token, to join the nodes to the cluster:

    ```
    $ TOKEN=YOUR_TOKEN
    ```

3. Now, if you need a multi-cluster configuration, execute the following command:

    ```
    $ curl -sfL https://get.k3s.io | INSTALL_K3S_EXEC="server
    --server https://MASTER_IP:6443" K3S_TOKEN=$TOKEN sh -s -
    ```

Now that you have learned how to install the `etcd` feature for K3s, let's move on to the next section to learn how to create and restore `etcd` snapshots for your Kubernetes objects configurations.

Creating and restoring etcd snapshots

K3s includes an experimental feature to back up and restore `etcd`. In this section, you are going to learn how to perform `etcd` snapshots and restoration for `etcd`. The snapshots are enabled by default with this backend. These snapshots are stored in `/var/lib/rancher/k3s/server/db/snapshots`. To create a backup, manually execute the following steps:

1. Create a backup manually:

    ```
    $ k3s etcd-snapshot --name=mysnapshot
    ```

    This will generate a file inside the `snapshots` folder.

2. To restore your `etcd` from this backup, execute the following command:

    ```
    $ k3s server \
    --cluster-reset \
    --cluster-reset-restore-path=<PATH-TO-SNAPSHOT>
    ```

3. You can automate the snapshot generation with the following option:

    ```
    --etcd-snapshot-schedule-cron
    ```

For more references to configure this, visit this link: `https://rancher.com/docs/k3s/latest/en/backup-restore/#options`.

You can even use the official documentation of `etcd`: `https://github.com/etcd-io/website/blob/main/content/en/docs/v3.5/op-guide/recovery.md`.

That's how you manage your `etcd` snapshots. Now, let's take a recap of what we have covered in this chapter.

Summary

This chapter covered common advanced configurations for Kubernetes edge clusters using Ubuntu and K3s. One of these common configurations was to install a bare metal load balancer using MetalLB. We also discussed the pros and cons of this as compared to the default K3s load balancer, KlipperLB, followed by the use cases of when to use each one. Then, we jumped to the advanced storage configurations of Longhorn, which is a really lightweight storage solution, and easy to install and configure for ReadWriteMany access modes for storage. Finally, we saw how to upgrade our cluster, and perform backups and restorations when using different data storage such as SQL or `etcd`. With all this knowledge, you are ready to create a production-ready cluster. In the next chapter, we are going to learn how to use k3OS to create your clusters using the K3s ISO image and overlay installation.

Questions

Here are a few questions to validate your new knowledge:

1. When do you choose KlipperLB or MetalLB as a bare metal load balancer solution?
2. How can I troubleshoot my MetalLB installation?
3. How can I install Longhorn to get more robust data storage solutions for my deployments?
4. How can I troubleshoot my Longhorn installation?
5. What other data storage solutions can I use instead of Longhorn?
6. What are the steps to upgrade my K3s clusters?
7. What are the steps to back up or restore my Kubernetes object configurations if I use a SQL backend or `etcd`?

Further reading

You can refer to the following references for more information on the topics covered in this chapter:

- What is the OSI Model?: `https://www.cloudflare.com/en-gb/learning/ddos/glossary/open-systems-interconnection-model-osi/`

- MetalLB official documentation: `https://metallb.universe.tf`

- MetalLB in layer 2 mode: `https://metallb.universe.tf/concepts/layer2`

- Kubernetes 101: Why You Need To Use MetalLB: `https://www.youtube.com/watch?v=Ytc24Y0YrXE`

- MetalLB ConfigMap configuration: `https://metallb.universe.tf/configuration`

- Persistent Volumes: `https://kubernetes.io/docs/concepts/storage/persistent-volumes`

- Volumes and Storage: `https://rancher.com/docs/k3s/latest/en/storage`

- Longhorn official page: `https://longhorn.io`

- Installing OpenEBS with RWM support: `https://docs.openebs.io/docs/next/rwm.html`

- Installing Rook with RWM support: `https://rook.io/docs/nfs/v1.7`

- Upgrading a K3s cluster: `https://rancher.com/docs/k3s/latest/en/upgrades`

- Backing up and restoring a K3s cluster: `https://rancher.com/docs/k3s/latest/en/backup-restore`

- Installation options: `https://rancher.com/docs/k3s/latest/en/installation/install-options/#registration-options-for-the-k3s-server`

4
k3OS Installation and Configurations

In edge computing contexts, companies are looking to simplify their tasks while using edge devices. Talking about the success of the K3s adoption, some industries need a ready-to-use operating system that can include an edge Kubernetes distribution. This is where k3OS fits the industry's needs. k3OS was designed to speed up the installation of K3s on edge devices.

k3OS packages all the necessary software to install K3s. This chapter explores how to use the k3OS ISO image to install K3s on x86_64 devices and how to use configuration files. You will learn from different configuration examples how to customize your K3s installations, from single node to multi-node. Finally, you will learn how to install k3OS on ARM devices using the overlay installation, taking in detailed configurations such as networking, the hostname that you would need when installing k3OS on edge devices, especially when you use ARM devices.

In this chapter, we're going to cover the following main topics:

- k3OS installation for x86_64 devices using an ISO image
- Advanced installations of k3OS using config files
- Multi-node ARM overlay installation

Technical requirements

For this chapter, you need one of the following VMs or devices:

- A Raspberry Pi 4B model with 4 GB RAM (minimum suggested)
- A cloud server or VM with **Ubuntu 20.04 LTS**
- A VM created using VirtualBox or other software for virtualization

In addition, you need a cloud storage service such as Amazon S3, Google Cloud Storage, or similar, to upload the configuration file for k3OS.

With this, we are ready to learn how to install k3OS in your preferred edge device. So, let's get started.

For more detail and code snippets, check out this resource on GitHub: `https://github.com/PacktPublishing/Edge-Computing-Systems-with-Kubernetes/tree/main/ch4`

k3OS – the Kubernetes operating system

k3OS is a Linux distribution that includes the minimal kernel, drivers, and binaries that you need to install Kubernetes at the edge. It features a lightweight distribution of Kubernetes called K3s. k3OS could be used as a fast operating system solution to install a lightweight Kubernetes cluster; this means that the k3OS image could be used to install master and agent nodes. k3OS uses K3s as the main software to create a single or multi-cluster node for the edge. So, with k3OS, you are ready to run your edge clusters without spending a lot of time.

k3OS can be installed using the following methods:

- **ISO image**
- **Overlay installation**

There are other methods to install a K3s cluster, but this is an easier way to get started fast. At this moment, k3OS is under development but supports a lot of features for ARM devices. So, let's move on to learn how to install your lightweight Kubernetes cluster using k3OS in the next section.

k3OS installation for x86_64 devices using an ISO image

We are going to install k3OS in a VM created using VirtualBox and an ISO image for a x86_64 architecture using a Macintosh. To do this, follow these steps:

1. Download the `k3OS x86_64 v0.20.7 ISO` image from this link:

    `https://github.com/rancher/k3os/releases`

2. Open your VirtualBox. The main window will appear, as shown in the following screenshot:

Figure 4.1 – VirtualBox main window

3. Enter the name to identify the VM – in this case, k3OS – and choose the type of VM as **Linux** and the **version** as **Other Linux (64 bit)**. Then, click on **Continue**.

Figure 4.2 – Name and operating system dialog

4. Choose at least **2048 MB** of RAM memory for the Live CD and interactive installation. Then, click on **Continue**.

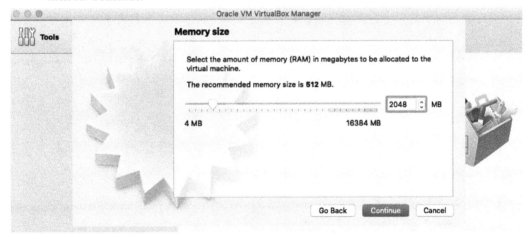

Figure 4.3 – Memory size dialog

5. Now let's create the hard disk; choose **Create a virtual hard disk now** and click on **Create**.

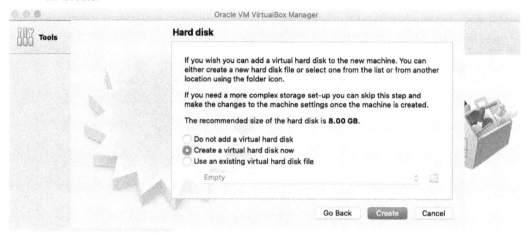

Figure 4.4 – Hard disk dialog

6. Choose **VDI** for **Hard disk file type** and click **Continue**.

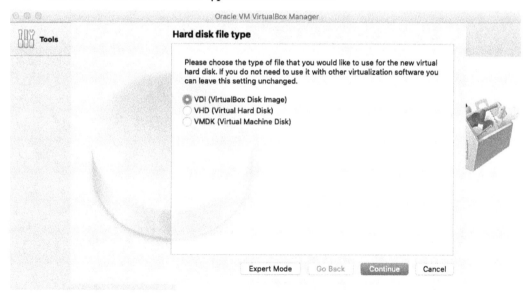

Figure 4.5 – Hard disk file type dialog

7. Choose **Dynamically allocated** for physical storage, as this will dynamically allocate the space for your hard disk. Then, click **Continue**.

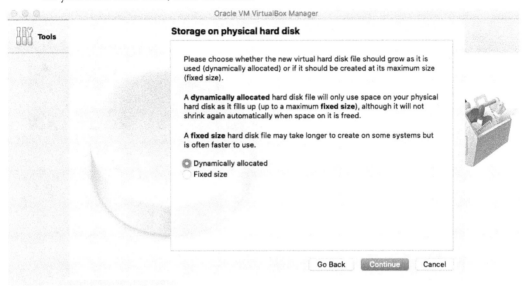

Figure 4.6 – Hard disk type dialog

8. Choose at least **4.00 GB** of disk space for your installation; it could be more, depending on your own requirements. Then, click on **Create**.

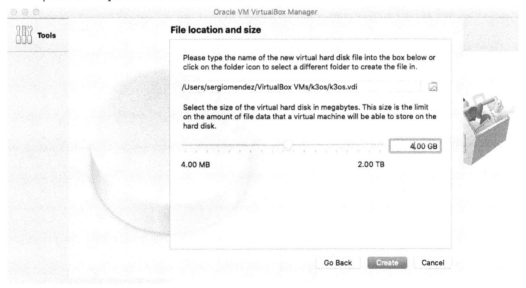

Figure 4.7 – Disk space dialog

9.   VirtualBox is going to move you to the **main window**; click on the **Settings** icon.

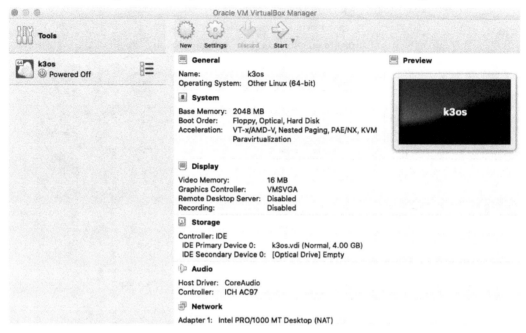

Figure 4.8 – Main window and Settings icon

10. Click on **Storage** and choose the **Empty** drive icon. Now, click on the small CD icon (⊙) next to the **Optical Drive** combo box.

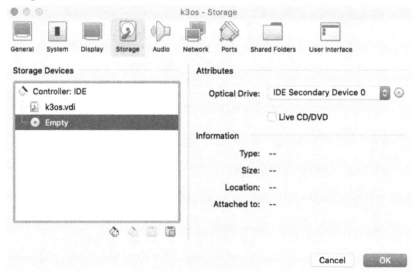

Figure 4.9 – Storage dialog

11. Choose the **Choose a disk file...** menu.

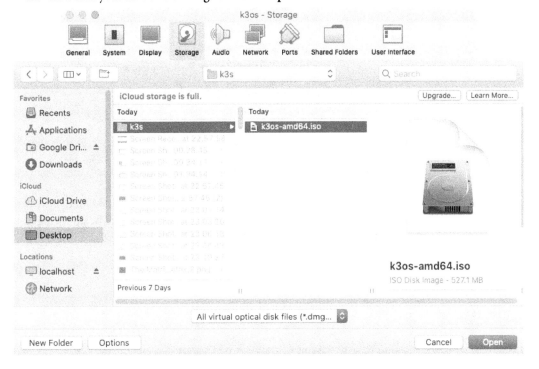

Figure 4.10 – Optical drive options

12. Now find your **k3OS ISO image** and click **Open**:

Figure 4.11 – Open ISO dialog

13. Click on the **Live CD/DVD** checkbox and then click **OK**.

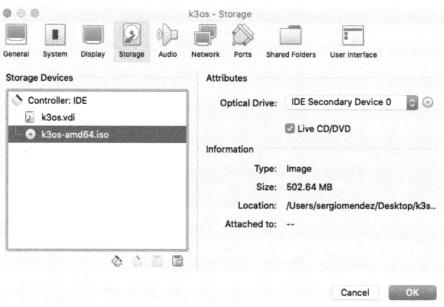

Figure 4.12 – Activating the Live CD/DVD feature

14. Now click on **Network**, and under **Adapter 1**, change the **Attached to**: combo box to **Bridged Adapter** for your VM to get an IP inside your local network. This is useful to access the VM remotely. Then, click **OK**.

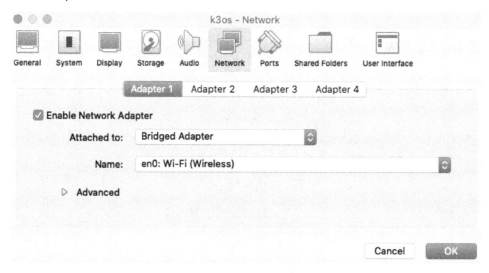

Figure 4.13 – Network configuration dialog

15. VirtualBox is going to return to the main window again; now click on the **Start** icon.

16. Now, you should wait for the process of loading the k3OS distribution. At the end, you are going to see a login prompt. Use `rancher` as both the username and password. For newer versions, you don't have to enter any password.

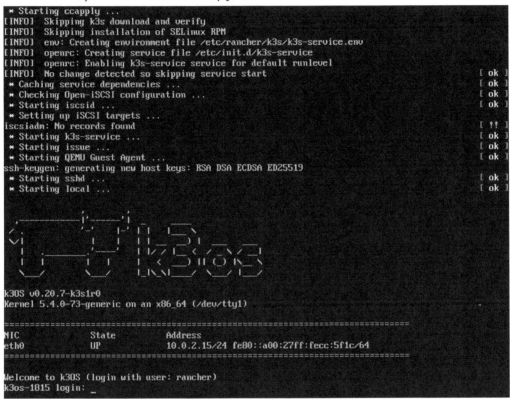

Figure 4.14 – k3OS live CD first-time login

17. Then, execute `sudo k3os install` to start the interactive script to install k3OS in your VM:

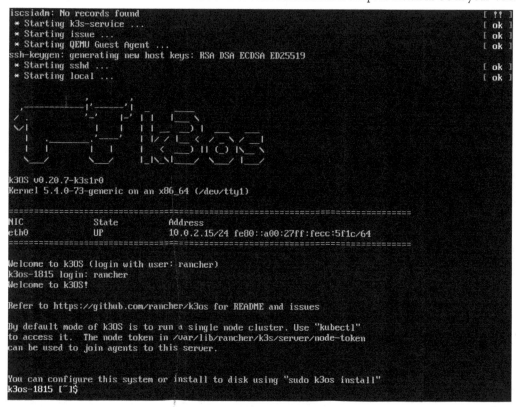

Figure 4.15 – k3OS starting the interactive installation

18. Complete the script installation with the following values:

- Install to disk with key 1.

- Config system with cloud-init with N.

- Authorize the GitHub user to SSH with N.

- Set up and confirm your new password for the rancher user with YOUR_PASSWORD.

- Configure Wi-Fi with N.

- Run as a server or agent with 1.

- Token or cluster secret – leave it empty and then press *Enter*.

- Your disk will be formatted with y.

The dialog will resemble the following screenshot:

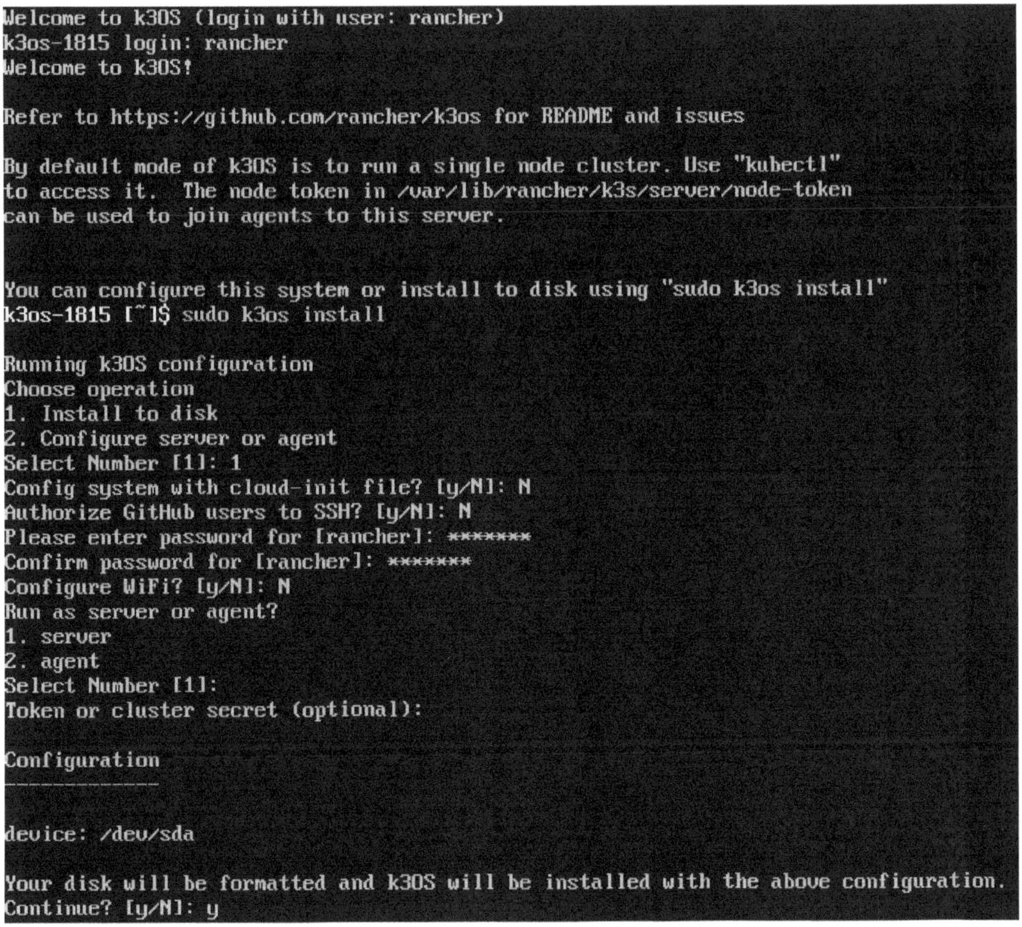

```
Welcome to k3OS (login with user: rancher)
k3os-1815 login: rancher
Welcome to k3OS!

Refer to https://github.com/rancher/k3os for README and issues

By default mode of k3OS is to run a single node cluster. Use "kubectl"
to access it.  The node token in /var/lib/rancher/k3s/server/node-token
can be used to join agents to this server.


You can configure this system or install to disk using "sudo k3os install"
k3os-1815 [~]$ sudo k3os install

Running k3OS configuration
Choose operation
1. Install to disk
2. Configure server or agent
Select Number [1]: 1
Config system with cloud-init file? [y/N]: N
Authorize GitHub users to SSH? [y/N]: N
Please enter password for [rancher]: *******
Confirm password for [rancher]: *******
Configure WiFi? [y/N]: N
Run as server or agent?
1. server
2. agent
Select Number [1]:
Token or cluster secret (optional):

Configuration
_____

device: /dev/sda

Your disk will be formatted and k3OS will be installed with the above configuration.
Continue? [y/N]: y
```

Figure 4.16 – Installation of the CLI dialog

19. After that, the VM is going to reboot; when starting to load, close the VM window, and the dialog from *Figure 4.17* will appear. Choose **Power off the machine** and then click **OK**.

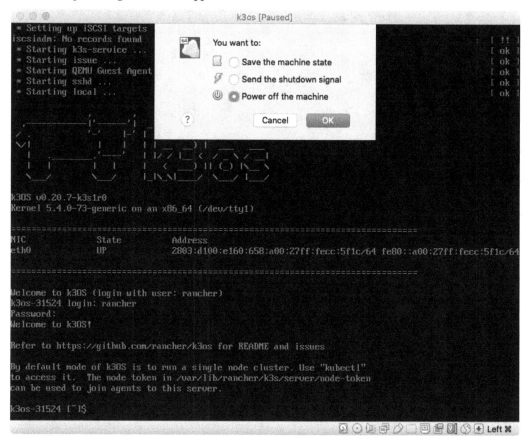

Figure 4.17 – Power off the VM

20. Once the VM is stopped, go to **Settings** and move to the **System** section under the **Motherboard** tab. Choose the **Boot order** section and select the **Hard Disk** option to prevent the optical disk from not loading again and prevent the launching of the installation script when the VM boots.

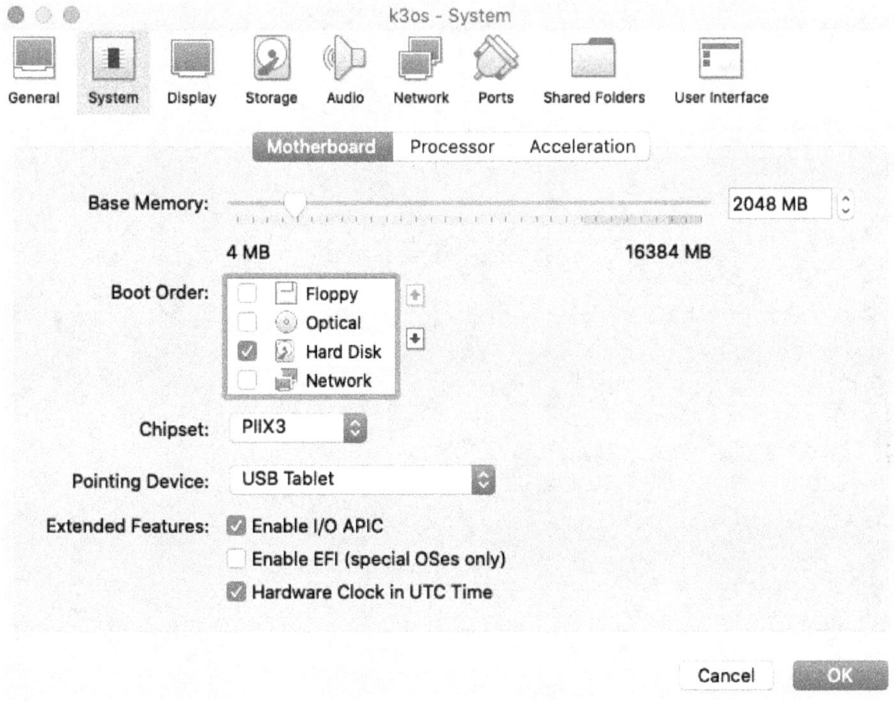

Figure 4.18 – Reconfiguring to boot on disk

21. Now, VirtualBox is going to return to the main window, so click on **Start** to start your VM again and your fresh k3OS installation will be loaded.

22. Now, your k3OS installation is running. You have to enter the `rancher` username and your new password to log in.

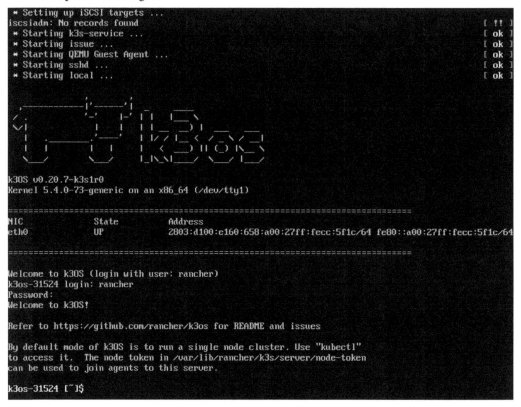

Figure 4.19 – k3OS first-time boot from disk

Now, you have a fresh installation of k3OS that is ready to use.

In the next section, you are going to learn how to do more advanced configurations using the same steps of creating a VM for k3OS, so let's move on to the next section.

Advanced installations of k3OS using config files

Now, we are ready to learn how to use config files to install k3OS; for this, you need a public GitHub repository where you can push these files. Before creating a `config.yaml` file to install k3OS, let's understand the different sections of this file. You are going to need a file for your master node and one for your agent node.

k3OS config file sections

Let's begin with an explanation of the sections to configure the host. These are as follows:

- **Hostname section:**

```
hostname: master
```

  Here is where you set the hostname – in this case, `master`.

- **SSH section:**

```
ssh_authorized_keys:
- ssh-rsa YOUR_KEY
```

  This section is used to set the default SSH public key that can access the node using SSH. You have to replace YOUR_KEY with your own public key.

- **Write files section:**

```
write_files:
  - path: /var/lib/connman/default.config
    content: |-
      [service_eth0]
      Type=ethernet
      IPv4=192.168.0.11/255.255.255.0/192.168.0.1
      IPv6=off
      Nameservers=8.8.8.8
```

  This section defines your network configuration. In this case, we have connected our Raspberry Pi to an ethernet connection with the internet. The IP of this node is set to `192.168.0.11`, the mask to `255.255.255.0`, and the gateway to `192.168.0.1`, and this connection is going to use the `8.8.8.8` nameserver. Remember that you can customize these values as per your internet provider.

Now, let's explore the sections to configure K3s. These are as follows:

- **k3OS DNS nameservers section:**

```
k3os:
  dns_nameservers:
  - 8.8.8.8
  - 1.1.1.1
```

This section sets the default DNS for the pods in the cluster; in this case, it is set to 8.8.8.8 and 1.1.1.1.

- **NTP servers section**:

```
ntp_servers:
- 0.us.pool.ntp.org
- 1.us.pool.ntp.org
```

This section sets the NTP servers to synchronize the time; in this case, it is set to 0.us.pool.ntp.org and 1.us.pool.ntp.org.

- **Password section**:

```
password: rancher
```

Here is where you set the password to access the host with k3OS.

- **Labels section**:

```
labels:
    region: america-central-1
```

This section configures the labels of the node; this is equivalent to using the kubectl labels command.

We have explained the common sections for the master and agent nodes. Now, let's continue with the sections that are different for the master and agent nodes.

Configurations for master and agent nodes

This section describes the specific sections that you have to use to configure a master or agent node. Let's get started with the master node:

- This is an example of a specific configuration for a master node:

```
k3os:
  token: myclustersecret
  password: rancher
  k3s_args:
  - server
  - "--write-kubeconfig-mode"
  - "644"
```

This k3OS configuration sets the master node, including the token that will be used to add new agent nodes, the password for the node, and parameters to send to the server binary. In this case, only modify the installation to execute `kubectl` without using `sudo`.

- This is an example of a specific configuration for an agent node:

```
k3os:
  server_url: https://192.168.0.11:6443
  token: myclustersecret
  password: rancher
```

This k3OS configuration sets an agent node to connect to the master node defined in `server_url`, using the token defined to be added to the cluster, and uses the defined password to access the node.

> **Important Note**
>
> For more options or arguments, you can check the next link: `https://rancher.com/docs/k3s`.

We have explained all the basic sections to create your cluster using config files. In the next section, we are going to create basic configuration files to create a multi-node cluster. So, let's get started.

Multi-node cluster creation using config files

Now is time to configure your cluster using config files, so let's put all the pieces together; the file for a master node will look like the following:

```
hostname: master
ssh_authorized_keys:
- ssh-rsa YOUR_PUBLIC_SSH_KEY
write_files:
  - path: /var/lib/connman/default.config
    content: |-
      [service_eth0]
      Type=ethernet
      IPv4=192.168.0.11/255.255.255.0/192.168.0.1
      IPv6=off
      Nameservers=8.8.8.8
k3os:
  dns_nameservers:
```

```
- 8.8.8.8
ntp_servers:
- 0.us.pool.ntp.org
password: rancher
token: myclustersecret
k3s_args:
- server
- "--write-kubeconfig-mode"
- "644
```

The configuration file for an agent node will look like this:

```
hostname: node01
ssh_authorized_keys:
- ssh-rsa YOUR_PUBLIC_SSH_KEY
write_files:
  - path: /var/lib/connman/default.config
    content: |-
      [service_eth0]
      Type=ethernet
      IPv4=192.168.0.12/255.255.255.0/192.168.0.1
      IPv6=off
      Nameservers=8.8.8.8
k3os:
  server_url: https://192.168.0.11:6443
  token: myclustersecret
  dns_nameservers:
  - 8.8.8.8
  ntp_servers:
  - 0.us.pool.ntp.org
  password: rancher
```

Now that we have the basic configuration files for master and agent nodes, it is time to use these files to deploy your multi-node cluster, as discussed in the following section.

Creating a multi-node K3s cluster using config files

Before you start creating the multi-node cluster, you must be equipped with the following requirements:

- 1 VM for the master node.

- 1 VM for the agent node.

- Your `master_example.yaml` master config file uploaded to a cloud storage service, such as Amazon S3, Google Cloud Storage, or similar. For example, if you use Google Storage, the URL for your file will be like this: `https://storage.googleapis.com/k3s/master_example.yaml`.

- Your `agent_example.yaml` agent config file uploaded to a cloud storage service, such as Amazon S3, Google Cloud Storage, or similar. For example, if you use Google Storage, the URL for your file will be like this: `https://storage.googleapis.com/k3s/agent_example.yaml`.

> **Important Note**
>
> The VM for the master and agent nodes must be configured to boot using the k3OS ISO image; in the next section, we will explain how to run the installation for master and agent nodes.

Now, we are ready to create our K3s cluster using config files. The next section explains how to use the k3OS ISO image to install a multi-node cluster. So, let's get started.

Creating a master node with config files

To create a master node, follow these steps:

1. Load your VM and then log in.

2. Log in to the VM with the `rancher` username and password.

3. Run the following command to start the k3OS installation:

   ```
   $ sudo k3os install
   ```

4. Follow the installation script using the following options:

   ```
   Choose Operation
   1. Install to disk
   2. Configure server or agent node
   Select number [1]: 1
   Config system with cloud-init file? [y/N] y
   cloud-init file location (file PATH or http URL):
   https://storage.googleapis.com/k3s/master_example.yaml
   ```

```
Your disk will be formatted and k3OS will be installed
with the above configuration.
Continue? [y/N] y
```

Remember to boot from the disk the next time to load your k3OS installation, following the last steps of the *k3OS installation for x86_64 devices using an ISO image* section. Now, let's move on to agent node creation.

Creating an agent node with config files

Now, follow the next steps to create an agent node:

1. Load your VM and then log in.

2. Log in to the VM with the `rancher` username and password.

3. Run the following command to start the k3OS installation:

   ```
   sudo k3os install
   ```

4. Follow the installation script using the following options:

   ```
   Choose Operation
   1. Install to disk
   2. Configure server or agent node
   Select number [1]: 1
   Config system with cloud-init file? [y/N] y
   cloud-init file location (file PATH or http URL):
   https://storage.googleapis.com/k3s/agent_example.yaml
   Your disk will be formatted and k3OS will be installed
   with the above configuration. Continue? [y/N] y
   ```

Remember to boot from the disk as you did with your master node. Now that your cluster is ready to be used, try to run `kubectl get nodes`. To verify whether your node was added, the output of the command should display the node name and the `Ready` status.

You have installed a multi-node cluster with master and agent nodes using VMs. Now it's time to install a multi-node cluster using ARM devices; in the next section, we are going to explore this kind of setup.

Multi-node ARM overlay installation

An overlay installation replaces some parts of your current OS installation or some parts of your system. In this case, when you use the **rootfs k3OS** file to perform this kind of installation, you will overwrite the `/sbin/init` file. Then, when you reboot your ARM device, the user space will be initialized and k3OS will be loaded. This kind of installation is supported for ARMv7 and ARM64 devices. One important thing is that you can customize this installation using the config YAML files, which must be stored on `/k3os/system/config.yaml`.

Before performing this overlay installation, you need the following:

- An ARMv7 or ARM64 device, such as a Raspberry PI with **Ubuntu 20.04 LTS** installed (you can use **Raspberry PI Imager** or **balenaEtcher**; check *Chapter 3, K3s Advanced Configurations and Management*, for reference)
- A network device connection with access to the internet and **Dynamic Host Configuration Protocol** (**DHCP**) to auto-assign an IP to your device
- An HDMI port connected to your monitor
- A keyboard connected
- Raspberry Pi Imager installed on your Macintosh or PC

In this case, we are going to use a Raspberry Pi 4B with 8 GB of RAM and a 64 GB Micro SD card for the master node, and 4 GB of RAM and a 32 GB Micro SD card for storage for the agent node. So, let's get started with the overlay installation for the master node first.

Master node overlay installation

Follow these steps to install the master node overlay:

1. Install Ubuntu 20.04 LTS with ARM64 by navigating to **Operating System | Other general purpose OS | Ubuntu | Ubuntu 20.04.2 LTS (RPi 3/4/400)**. Insert your Micro SD card and choose it in **Storage**; then click on **WRITE**, as shown in the following screenshot:

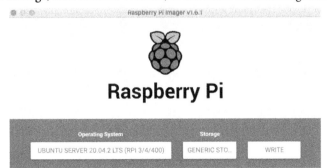

Figure 4.20 – Ubuntu server installation using Raspberry Pi Imager

This is going to ask for your credentials to start the installation; when the process finishes, it will show the following screen:

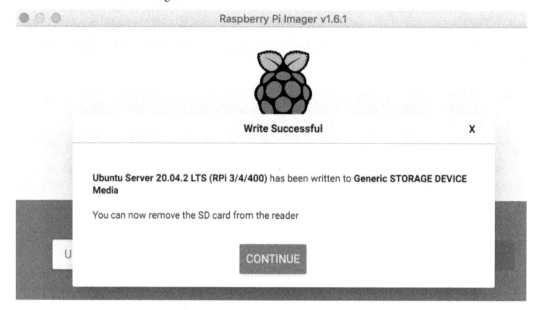

Figure 4.21 – Write Successful dialog

Now, extract your MicroSD card and insert it into your SD slot in your Raspberry Pi.

2. Turn on your Raspberry Pi.

3. Log in to your ARM device. The default user/password for Ubuntu is ubuntu; the system will ask you to set a new password. Set your desired password.

4. Install the net-tools package to see which IP has your ARM device with the following command:

    ```
    $ sudo apt-get update
    $ sudo apt-get install -y net-tools
    ```

5. Check for your IP with the following command:

    ```
    $ sudo ifconfig
    ```

6. Log in to your device using SSH, the ubuntu username, and the password that you set when logging in to your Ubuntu device for the first time. DEVICE_IP is the IP that ifconfig returned. Run this on your laptop to log in:

    ```
    $ ssh ubuntu@DEVICE_IP
    ```

7. Add the `cgroup_memory=1 cgroup_enable=memory` options to your kernel parameters in the `/boot/firmware/cmdline.txt` file with the following command:

```
$ sed 's/$/cgroup_memory=1 cgroup_enable=memory/' /boot/
firmware/cmdline.txt | sudo tee /boot/firmware/cmdline.
txt
```

Alternatively, you can use any editor to add this text at the end of the `cmdline.txt` file.

8. Run the following commands to install k3OS; this will unpack the `tar.gz` file into the `/` folder:

```
$ curl -sfL https://github.com/rancher/k3os/releases/
download/v0.20.7-k3s1r0/k3os-rootfs-arm.tar.gz | sudo tar
zxvf - --strip-components=1 -C /
```

If you want to configure a master node, follow these steps:

1. Download or copy a config into your local directory; let's use our previous master configuration using `wget`, plus the URL:

```
$ wget https://storage.googleapis.com/k3s/master_example.
yaml
```

2. Copy the configuration to `/k3os/system/config` using the following command:

```
$ sudo cp master_example.yaml /k3os/system/config.yaml
```

You can customize these files as you need; remember that this example uses an SSH key that you can use to log in to your node remotely.

(*Optional*) If you want to configure an **agent node**, follow the next steps:

1. Download or copy a config in your local directory, let's use our previous master configuration using `wget` plus the URL:

```
$ wget https://storage.googleapis.com/k3s/agent_example.
yaml
```

2. Copy the configuration to `/k3os/system/config` using the following command:

```
$ sudo cp agent_example.yaml /k3os/system/config.yaml
```

You can customize these files as you need; remember that this example uses an SSH key that you can use to log in to your node remotely.

3. Sync the filesystem for each node with the following command:

```
$ sudo sync
```

4.   Reboot the system:

```
$ sudo reboot -f
```

5.   Log in to your device:

```
$ sudo rancher@DEVICE_IP
```

Remember that this configuration has set a static IP for your node.

6.   If you are logged in to the master node, you can run the following command to check whether all the nodes were detected:

```
$ kubectl get nodes
```

Now, your K3s cluster has been installed and is ready to be used. You now know how to configure the cluster using the overlay installation, which is quicker compared to the execution of the default K3s script found on the K3s official website.

Summary

In this chapter, we learned how to install K3s using the k3OS, a production-ready Linux distribution, covering how to prepare your VMs in case that you want to create a cluster for x86_64 architectures. Then, we moved on to explain how configuration files are used to perform advanced and custom cluster installations, and how you can configure them to create a multi-node cluster using the ISO image or the overlay installation. Finally, we covered how to create a multi-node cluster using the overlay installation, to reduce the manual configurations to install K3s using the k3OS potential. Now, we are close to starting use cases and real configuration in the coming chapters. In the next chapter, we are going to create a production-ready cluster using all the things that we learned in the previous chapters.

Questions

Here are a few questions to validate your new knowledge:

1. What kind of k3OS installations are available if you are using x86 or x86_64 devices?
2. How do you install k3OS using an ISO image?
3. What kind of k3OS installations are available if you are using ARMv7 or ARM64 devices?
4. How do you install k3OS using overlay installation?
5. How can I use configuration files to customize my cluster installations?
6. How can I send parameters to my master or agent node using the k3OS arguments section?
7. How can I create a multi-node cluster using the k3OS ISO image or the overlay installation?
8. Where can I find more information about available parameters for K3s?
9. What are the other types of installations available for k3OS?

Further reading

You can refer to the following references for more information on the topics covered in this chapter:

- Official k3OS documentation: `https://github.com/rancher/k3os`
- Available releases for k3OS: `https://github.com/rancher/k3os/releases`
- K3s installation options to add custom parameters to your config files: `https://rancher.com/docs/k3s/latest/en/installation/install-options`
- k3OS image generator: `https://github.com/sgielen/picl-k3OS-image-generator`
- *Installing k3s with Alpine Linux on Raspberry Pi 3B+*: `https://blog.jiayihu.net/install-k3s-with-alpine-linux-on-raspberry-pi-3b`
- k3OS configuration file examples: `https://www.chriswoolum.dev/k3s-cluster-on-raspberry-pi`
- *How to launch ARM aarch64 VM with QEMU from scratch*: `https://futurewei-cloud.github.io/ARM-Datacenter/qemu/how-to-launch-aarch64-vm`
- k3sup for K3s cluster creation: `https://blog.alexellis.io/test-drive-k3s-on-raspberry-pi`

K3s Homelab for Edge Computing Experiments

At this point, we have explored essential topics to create your own edge computing cluster. The previous chapters covered how to configure and install a K3s cluster. Building small and big solutions at home involves experimenting. In this chapter, we are going to start building a simple but real cluster, using the knowledge acquired in the previous chapters. We will refer to this environment as the K3s homelab. Once this cluster is created, we are going to deploy a simple application. We will use this as a quickstart method of using Kubernetes with your cluster. In the last part of this chapter, we are going to use the Kubernetes dashboard as a simple UI to manage Kubernetes clusters.

In this chapter, we're going to cover the following main topics:

- Installing a multi-node K3s cluster on your local network

- Deploying your first application with `kubectl`

- Deploying a simple NGINX server using YAML files

- Adding persistence to your applications

- Deploying a Kubernetes dashboard

Technical requirements

For this chapter, you need the following hardware to create your K3s homelab for your edge computing applications or experiments:

- Two or more Raspberry Pi 4 B models with a minimum of 4 GB RAM and a 32 GB microSD card with Ubuntu version 20.04 or later. The SanDisk Extreme microSDHC 32 GB UHS-1 A1 V30 or similar is recommended as the microSD card.

- Ethernet cables to connect your Raspberries.

- An Ethernet internet connection for the Raspberries with **Dynamic Host Configuration Protocol (DHCP)** activated.

- One switch to connect your Raspberry to your local network.

With this hardware, we are ready to start building our K3s homelab. So, let's get started.

For more detail and code snippets, check out this resource on GitHub: `https://github.com/PacktPublishing/Edge-Computing-Systems-with-Kubernetes/tree/main/ch5`

Installing a multi-node K3s cluster on your local network

To start creating this homelab, let's understand the network topology that we are going to use. Each component in the following diagram is used in the homelab:

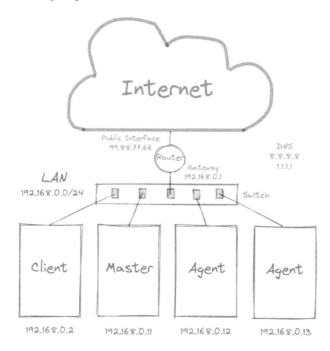

Figure 5.1 – Homelab architecture

Here is a small explanation of each component in the figure:

- **Lan**: This is the local network that you are going to use. In our example, the network is defined as `192.168.0.0/24`.

- **Switch**: A switch is also a network connecting device that connects various devices in the same network.

- **Router**: A router connects devices across multiple networks. Typically, home routers are hybrid devices that give local computers access to the internet. It also has small switch capabilities to connect local computers, using wireless or Ethernet ports for your wired devices.

- **Public Interface**: This is the interface of your router that has a public IP.

- **Gateway**: This is an IP address that is used as the gateway in your private network.

- **DNS**: This is the DNS IP address that your router is going to assign to your devices automatically – in this case, `8.8.8.8` and `1.1.1.1`.

- **Master**: This is the master node of your K3s cluster.

- **Agent**: This is the agent node that acts as a worker in your K3s cluster.

- **Client**: This is your local machine or laptop where you are going to access the cluster using the `kubectl` command.

Now – a small explanation about how these pieces interact with each other. All your machines will use the `192.168.0.0/24` network; in this case, let's think that your client will use the `192.168.0.2` IP. Using the config files or parameters to install your cluster, you can choose an IP range inside the previous network for your nodes. In this case, the master is using the `192.168.0.11` IP and your agents are using the `192.168.0.12` and `192.168.0.13` IP addresses. Remember that your configuration has set static IP private addresses to your nodes to prevent errors in your nodes if the IP address changes. We assume that the nodes are using IP addresses starting from `192.168.0.11` to `192.168.0.13`. We are going to use the `192.168.0.240` to `192.168.0.250` IP address range for load balancers. This is just a simple example of how to organize your IPs for your cluster.

We are assuming that your router is in the `192.168.0.0/24` network. As we mentioned, home routers have some switch capabilities to auto-assign dynamic IP addresses using a DHCP service configured inside the router, but this isn't healthy for your nodes. That's the main reason for using static IPs for your nodes. We are assuming some public IP to use as an example. We are assuming that we are going to use the `8.8.8.8` and `1.1.1.1` DNS servers.

Important Note

All these IP ranges can change, depending on your internet provider or the router device that you are using. We set these values to provide an example of how to organize the network for your cluster.

To create your homelab, we have to complete the following tasks:

1. Install Ubuntu image on your Raspberry device.

2. Configure your device to run the K3s installer.

3. Configure the K3s master node.

4.   Configure the K3s agent nodes.

5.   Install **MetalLB** as the load balancing service.

6.   Install **Longhorn** as the default storage class.

7.   Configure `kubectl` in an external client to access the cluster.

8.   Deploy your first application using `kubectl` and YAML files.

9.   Install and configure **Lens** to manage your cluster.

So, let's now quickly recap the concepts, starting with how to install Ubuntu on your device.

Installing an Ubuntu image on your Raspberry device

In this section, we are going to install an Ubuntu image on a Raspberry device. You can skip this section or refer to previous chapters for more information. As a quick summary, you can follow the next steps to install Ubuntu on a Raspberry device:

1.   Open *Raspberry Pi Imager*.

2.   Click on the **CHOOSE OS** button to choose the Ubuntu Server 20.04 64-bit for ARM64 operating system, which is located in the **Other general purpose OS | Ubuntu** menu.

3.   Then, insert your microSD card (you may have to buy an adapter to read microSD cards); your device appears when you select the **CHOOSE STORAGE** button.

4.   Click on the **WRITE** button.

5.   Accept to write the device; then, Raspberry Pi Imager will ask you for your username and password in order to continue writing to the microSD card.

6.   Wait until the writing and verifying process finishes.

7.   Extract your microSD card.

8.   Insert the microSD card into your Raspberry Pi and turn it on.

9.   Repeat these steps for each Raspberry Pi device that will be part of your cluster.

Now, let's move to configure the network settings and the container support for your device.

Configuring your Raspberry Pi to run the K3s installer

In this section, we are going to configure the network settings, including your static IP address, DNS, hostname, and hosts files, finalizing with activating the support of the cgroups necessary to use **containerd**. Now, follow the next steps to perform the final setup before installing K3s in your nodes; remember that you can customize all these configurations to fit your own network:

1. Turn on your device.

2. When Ubuntu asks you for your username and password, enter the username and `ubuntu` as the password this is the default password for the first login.

3. Now, Ubuntu will ask you to change the default password; let's use `k3s123-` as our password.

4. Now, let's configure the network; by default, Ubuntu uses `cloud-init` to configure the network. Let's deactivate this by creating the `99-disable-network-config.cfg` file with the following commands and content:

    $ sudo nano /etc/cloud/cloud.cfg.d/99-disable-network-config.cfg

    Here is an example of the content:

    network: {config: disabled}

5. If you execute `ifconfig`, you will see that your device is `eth0`, but it can be named `es3` or something similar, so let's modify the `50-cloud-init` file with the following command:

    $ sudo nano /etc/netplan/50-cloud-init.yaml

6. Then, modify the content of the file; it has to look something like this:

    ```
    network:
      version: 2
      renderer: networkd
      ethernets:
        eth0:
          dhcp4: no
          addresses:
            - 192.168.0.11/24
          gateway4: 192.168.0.1
          nameservers:
              addresses: [8.8.8.8, 1.1.1.1]
    ```

7. Now, apply the configuration and reboot your device to see whether your IP address is set when the **Operating System (OS)** starts. To do this, execute the following command:

    ```
    $ sudo netplan apply
    ```

8. Now, configure the kernel parameters for the boot by editing the /boot/firmware/cmdline.txt file with the following command and content:

    ```
    $ sudo nano /boot/firmware/cmdline.txt
    ```

    Add this content to the end of the line:

    ```
    cgroup_memory=1 cgroup_enable=memory
    ```

9. Edit /etc/hostname using the master name for your master node. Use node01 and node02 for the hostnames of your agent nodes; let's edit the file using nano:

    ```
    $ sudo nano /etc/hostname
    ```

    Here is an example of the content:

    ```
    master
    ```

10. Edit the /etc/hosts file, adding the hostname; at a minimum, you need to have a line like this:

    ```
    $ sudo nano /etc/hosts
    ```

    Here is an example of the content:

    ```
    127.0.0.1 localhost master
    ```

> **Important Note**
> You can also use master.local instead of master to follow **Internet Engineering Task Force (IETF)** naming conventions for local networks. This may also help with zero-configuration **multicast DNS (mDNS)** setups. For more information, you can check out this link: http://www.zeroconf.org.

Now, reboot your device:

```
$ sudo reboot
```

This configuration is required to prepare your device to configure a K3s master or agent nodes. You can also follow IETF recommendations for local network design. In the next section, you will see how to install K3s for your master nodes.

Configuring the K3s master node

This section explains how to install your master node for your K3s cluster; for this, you have to follow these steps:

1. Turn on your device and log in with your ubuntu user.

2. Run the following commands to install your master node using MASTER_IP as 192.168.0.11, as shown in *Figure 5.1*, for your K3s cluster:

    ```
    $ MASTER_IP=<YOUR_PRIVATE_IP>
    $ curl -sfL https://get.k3s.io | INSTALL_K3S_EXEC="--
    write-kubeconfig-mode 644 --no-deploy traefik --disable
    traefik --tls-san "$MASTER_IP" --node-external-ip
    "$MASTER_IP" --disable servicelb" sh -s -
    ```

Now, we have installed the master node. This will be the node with the 192.168.0.11 IP address. Now, let's go ahead and add agent nodes to the cluster in the next section.

Configuring the K3s agent nodes

This section explains how to complete our initial cluster diagram by repeating this section twice to complete the configuration of two agent nodes. Agent nodes will use the 192.168.0.12 and 192.168.0.13 IP addresses. Complete the following steps to configure each agent node:

1. Log in to your master node:

    ```
    $ ssh ubuntu@<MASTER_IP>
    ```

    We are going to extract the servicer node token to connect the agent nodes. In this case, the master node will be 192.168.0.11.

2. Extract and copy the token to join your agent nodes in the cluster, running the following command:

    ```
    $ sudo cat /var/lib/rancher/k3s/server/node-token
    ```

3. Log out from your master node. Now, you have the token to join additional nodes to the cluster.

For each agent node to join the cluster, follow the next steps (the easy way):

1. Log in to your agent node that you want to add to the cluster. In this case, AGENT_IP will be 192.168.0.12 or 192.168.0.13:

    ```
    $ ssh ubuntu@<AGENT_IP>
    ```

2. Set an environment variable with the token that your master generated:

```
$ export TOKEN=<YOUR_MASTER_TOKEN>
```

3. Register your node with the following command; in this case, MASTER_IP will be
    192.168.0.11:

```
$ curl -sfL https://get.k3s.io | sh -s - agent --server
https://MASTER_IP:6443 --token ${TOKEN}
```

Exit from your agent node:

```
$ exit
```

Now, we have configured our agent nodes. Let's install MetalLB to start using load balancers for our applications.

Installing MetalLB as the load balancing service

MetalLB is a bare metal load balancer that can help when using the load balancing service of a regular Kubernetes cluster, with the capabilities of networking designed for bare metal, such as IP address assignment. So, let's get started by installing MetalLB by following these steps:

1. Create a MetalLB namespace (metallb-system) with the official manifests, executing the following lines:

```
$ kubectl apply -f https://raw.githubusercontent.com/
metallb/metallb/v0.10.2/manifests/namespace.yaml
```

2. Before running the command to install MetalLB, you have to create a ConfigMap resource called metallb-config inside the metallb-system namespace. Let's call this file config.yaml, with the following content:

```
apiVersion: v1
kind: ConfigMap
metadata:
  namespace: metallb-system
  name: config
data:
  config: |
    address-pools:
    - name: default
      protocol: layer2
```

```
addresses:
- 192.168.0.240-192.168.0.250
```

3. Now, create `ConfigMap`, executing the following command:

   ```
   $ kubectl apply -f config.yaml
   ```

4. Install MetalLB with the official manifests by executing the following lines:

   ```
   $ kubectl apply -f https://raw.githubusercontent.com/
   metallb/metallb/v0.10.2/manifests/metallb.yaml
   ```

Now, you have installed MetalLB. You are ready to install services that use load balancers. These load balancers are commonly found in a lot of Kubernetes software. Now, it is time to add Longhorn for our storage.

Installing Longhorn with ReadWriteMany mode

K3s includes basic storage support. Sometimes, this storage can cause errors when you are installing software. To prevent this, you will need another storage driver instead of the default one that comes with K3s. In this case, you can use Longhorn. With Longhorn, you can install Kubernetes software that looks for regular storage drivers. So, let's install Longhorn in the following steps:

1. Log in to your **virtual machine** (**VM**) or device:

   ```
   $ ssh ubuntu@NODE_IP
   ```

2. If you want to install **ReadWriteMany Persistent Volume** (**PVC**) mode, you have to install `nfs-common` on each VM with Ubuntu installed in your cluster. To do this, execute the following command:

   ```
   $ sudo apt install -y nfs-common
   ```

3. Apply the official Longhorn manifests, as follows:

   ```
   $ kubectl apply -f https://raw.githubusercontent.com/
   longhorn/longhorn/v1.1.2/deploy/longhorn.yaml
   ```

Now, you have Longhorn installed and running. Let's move on to learn how to configure `kubectl` on your personal computer to manage your K3s.

Extracting the K3s kubeconfig file to access your cluster

Now, it's time to configure the `kubeconfig` file to access your K3s cluster from your computer, using the `kubectl` command. To configure the connection of your new K3s cluster from the outside, follow these steps:

1. Install `kubectl`, following the instructions of the official documentation of Kubernetes (`https://kubernetes.io/docs`); in this case, we are going to use the instructions for Macintosh:

    ```
    $ curl -LO "https://dl.k8s.io/release/$(curl -L -s
    https://dl.k8s.io/release/stable.txt)/bin/darwin/amd64/
    kubectl"
    $ chmod +x ./kubectl
    $ sudo mv ./kubectl /usr/local/bin/kubectl
    $ sudo chown root: /usr/local/bin/kubectl
    ```

    Or you can install `kubectl` using `brew` on macOS, using the next command:

    ```
    $ brew install kubectl
    ```

    For other custom installations, such as `kubectl` for Apple's new silicon processors, Linux, or Windows, visit the Kubernetes official documentation: `https://kubernetes.io/docs/tasks/tools/install-kubectl-macos`.

2. From the master node, copy the content inside `/etc/rancher/k3s/k3s.yaml` to your local `~/.kube/config` file.

3. Change the permissions of the file with the next command:

    ```
    $ chmod 0400 ~/.kube/config
    ```

4. Change part of the server value from `127.0.0.1` to the `MASTER_IP` address of your master node; in this case, it will be `192.168.0.11`:

    ```
    server: https://127.0.0.1:6443
    ```

    This changes to the following:

    ```
    server: https://MASTER_IP:6443
    ```

> **Important Note**
>
> Remember to install `kubectl` before you copy the Rancher `kubeconfig` file onto your computer. Remember that the content of the `k3s.yaml` file has to be stored inside `~/.kube/config` and needs the `0400` permissions. To check how to install the `kubectl` command, go to `https://kubernetes.io/docs/tasks/tools/install-kubectl-macos`.

Now, we are ready to use the cluster. In the next section, we are going to deploy a basic application with `kubectl` and YAML files, using MetalLB and Longhorn. So, let's start deploying applications, using `kubectl` in the next section.

Deploying your first application with kubectl

This section covers the basics of Kubernetes. We are going to deploy an application using `kubectl` first. But before that, let me give you a quick introduction about how Kubernetes works with its basic objects.

Basic Kubernetes objects

Kubernetes works with objects that provide different functionalities for your application using containers. The goal of Kubernetes is to orchestrate your containers. Kubernetes uses two ways to create objects. One is using imperative commands – in the case of Kubernetes, the `kubectl` command. The other is using declarative files, where the state of an object is defined, and Kubernetes ensures that this state stays as it was defined throughout its lifetime:

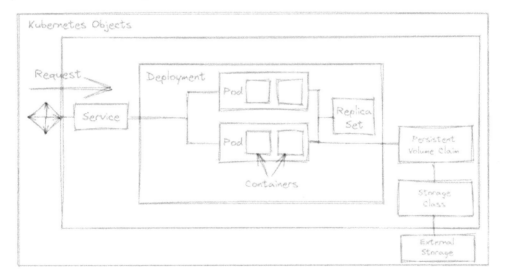

Figure 5.2 – Kubernetes objects

This diagram represents how some of the basic objects interact with each other to deploy and manage an application. So, let's explain each of these objects:

- **Pod** contains one or more containers, where your application lives; all the containers inside a Pod share the same network, memory, and CPU.

- **ReplicaSet** controls the number of pods to be the same.

- **Deployment** is an advanced kind of **ReplicaSet** object that not only controls the number of Pods and versions but also the changes of the Pods, providing a way to perform rollbacks.

- **Service** is a way to expose your services. There are different types. **NodePort** opens a random port on all the nodes, **ClusterIP** creates a DNS that you can use to communicate with your Pod or deploy with other Pods or deployments, and **LoadBalancer** creates an exclusive endpoint to publish your app to the outside.

- **Persistent Volume Claim** is the object in charge of requesting persistent storage and creating stateful deployments.

- **Storage Class** is the object that defines how you are going to request storage for an application.

With these pretty basic concepts, let's move on to the practical aspects to understand how each component works. In the next section, we are going to deploy a simple NGINX server using `kubectl`.

Deploying a simple NGINX server with pods using kubectl

In this section, we are going to deploy an NGINX server, step by step, using `kubectl`. To do this, follow these steps:

1. Create a pod with the `nginx` image:

    ```
    $ kubectl run myserver --image=nginx --restart=Never
    ```

2. Create a `LoadBalancer` type of service for this Pod to expose and access the NGINX pod:

    ```
    $ kubectl expose pod/myserver --port=8001 --target-
    port=80 -type=LoadBalancer
    ```

3. Assign the IP address to your load balancer with the following command:

    ```
    $ IP_SERVICE=$(kubectl get svc mywebserver --output
    jsonpath='{.status.loadBalancer.ingress[0].ip}')
    ```

4. Access the next URL using your browser or the following command:

    ```
    $ curl IP_SERVICE:8001
    ```

Now, you have an NGINX service up and running. So, let's move to deploy a **Redis** database that you can access to store data in the next section.

Deploying a Redis NoSQL database with pods

Now, we are going to deploy a Redis NoSQL key-value database that you can access to store some data. We chose Redis as a basic example as it is quick and easy to use. So, let's deploy Redis using the following commands:

1. Create a pod with a `redis` image:

    ```
    $ kubectl run myredis --image=redis --restart=Never
    ```

2. Create a `ClusterIP` service that you can use to connect to Redis using the name of the service:

    ```
    $ kubectl expose pod myredis --port=6379 --type=ClusterIP
    ```

3. Let's create an `ubuntu` client with the next command:

    ```
    $ kubectl run client -it --rm --image=ubuntu:18.04 --
    bash
    ```

4. Now, you are inside the client, so let's install the Redis client to get connected to the Redis pods with the following command:

    ```
    root@client# apt-get update;apt-get install -y redis-
    tools
    ```

5. Store the variable with the value 1 and get the value from the client, using the following commands:

    ```
    root@client# redis-cli -h myredis set a 1
    root@client# redis-cli -h myredis get a
    ```

    The last command returns the value of the a variable, which is 1.

6. Write `exit` and then press *Enter* to exit the client. The client will be automatically deleted because of the `--rm` parameter.

7. Now, let's expose Redis, using `NodePort` as an example of how to expose a pod using the IPs of your nodes:

    ```
    $ kubectl expose pod myredis --name=myredis-nodeport
    --port=6379 --type=NodePort
    ```

Now, you can access your Redis database using the IP of the host where Redis was deployed.

You have finished installing a simple database – in this case, Redis. Now, let's explore the deployment objects and storage in the next section.

Deploying and scaling an NGINX server with deployments

One of the advantages of using deployments is that you manage the changes of your deployment if the version or the configuration changes. Let's deploy a simple NGINX server, scale the deployment, change the image, and then perform a rollback to see the power of deployments. Deploy the NGINX server by following these steps:

1. Create a deployment with two replicas using the `nginx` image:

   ```
   $ kubectl create deployment mywebserver --image=nginx
   --replicas=2
   ```

2. Create a `LoadBalancer` service to expose your deployment:

   ```
   $ kubectl expose deployment mywebserver --port=8002
   --target-port=80 --type=LoadBalancer
   ```

3. Create the IP for `mywebserver`:

   ```
   $ IP_SERVICE=$(kubectl get svc mywebserver --output
   jsonpath='{.status.loadBalancer.ingress[0].ip}')
   ```

4. Access the web server using `curl`:

   ```
   $ curl $IP_SERVICE:8002
   ```

5. Scale `mywebserver` with 0 replicas:

   ```
   $ kubectl scale deploy/mywebserver --replicas=0
   ```

6. Try to access `mywebserver` again:

   ```
   $ curl $IP_SERVICE:8002
   ```

7. Scale `mywebserver` with two replicas and wait until the deployment is ready; you can check this with the following:

   ```
   $ kubectl scale deploy/mywebserver --replicas=2
   $ kubectl rollout status deploy/mywebserver
   ```

8. Try to access mywebserver again:

   ```
   $ curl $IP_SERVICE:8002
   ```

9. Let's change the nginx version of the deployment with the wrong version:

   ```
   $ kubectl set image deployment/mywebserver
   nginx=nginx:1.16.1.x
   ```

10. Check the changes in the description of the object:

    ```
    $ kubectl describe deployment mywebserver | grep -i image
    ```

11. Check the current pod status for the mywebserver deployment:

    ```
    $ kubectl get pods
    ```

    You will see some pods from mywebserver with errors.

12. Let's roll back to the previous version:

    ```
    $ kubectl rollout undo deploy/mywebserver
    ```

13. Check the current pod status for the mywebserver deployment:

    ```
    $ kubectl get pods
    ```

    You will see that the pods with errors have disappeared because you returned to the previous image that the deployment was using – in this case, the correct image name.

Now, you have deployed your application using the deployment object. Let's do something similar using YAML files and add some persistence. To do this, let's move on to the next section.

Deploying a simple NGINX server using YAML files

At this point, our examples don't store data and the objects are created using imperative commands. To use declarative files, you can use the kubectl command to generate the files. Remember to deploy your application, using pods or deployments – just choose one of these options. To start, let's create an NGINX pod using YAML files.

Deploying an NGINX server using a Pod

Now, let's create an NGINX pod using YAML files. To do this, follow these steps:

1. If you want to use pods, you can use the next YAML file. To generate the file, use the following command:

    ```
    $ kubectl run nginx --image=nginx --dry-run -o yaml >
    nginx-pod.yaml
    ```

    The nginx-pod.yaml file will look like this:

    ```
    apiVersion: v1
    kind: Pod
    metadata:
      name: nginx
      labels:
        name: nginx
    spec:
      containers:
      - name: nginx
        image: nginx
        ports:
          - containerPort: 80
    ```

2. Apply the generated file using the following command:

    ```
    $ kubectl create -f nginx-pod.yaml
    ```

Let's move on to create an NGINX deployment in the next section.

Deploying an NGINX server using deployment

So, let's get started creating an NGINX server using deployment with YAML files. To do this, follow these steps:

1. Generate the YAML file for deployment using the following command:

    ```
    $ kubectl create deployment nginx --image=nginx
    --replicas=2 --dry-run -o yaml > nginx-deployment.yaml
    ```

The file `nginx-deployment.yaml` will look like this:

```
apiVersion: apps/v1
kind: Deployment
metadata:
  labels:
    app: nginx
  name: nginx
spec:
  replicas: 2
  selector:
    matchLabels:
      app: nginx
  template:
    metadata:
      labels:
        app: nginx
    spec:
      containers:
      - image: nginx
        name: nginx
```

2. Apply the generated file using the following command:

```
$ kubectl create -f nginx-deployment.yaml
```

Now that we have learned how to create a pod and deployment in Kubernetes, let's move on to the next section to expose these objects using services with YAML files.

Exposing your pods using the ClusterIP service and YAML files

To communicate your pod or deployment with other applications, you may need a DNS record. The ClusterIP service type creates a DNS A record for your pod or deployment. Using this DNS, other objects in your cluster can access your application. So, let's create a ClusterIP service for your application, following these steps:

1. To expose your application using YAML files, generate the YAML file for the ClusterIP service type:

    ```
    $ kubectl expose pod/nginx --type=ClusterIP --port=80
    --target-port=8001 --dry-run -o yaml > nginx-clusterip.
    yaml
    ```

    The nginx-service.yaml file will look like this:

    ```
    apiVersion: v1
    kind: Service
    metadata:
      creationTimestamp: null
      labels:
        app: nginx
      name: nginx
    spec:
      ports:
      - port: 80
        protocol: TCP
        targetPort: 8001
      selector:
        app: nginx
      type: ClusterIP
    ```

2. Apply the generated file using the following command:

    ```
    $ kubectl create -f nginx-pod.yaml
    ```

Now that you have learned how to create a ClusterIP service using YAML files, let's move on to creating a NodePort service for your application in the next section.

Exposing your pods using the NodePort service and YAML files

To create a NodePort service for a previously created pod, follow these steps:

1. For NodePort, use the following command:

    ```
    kubectl expose pod/nginx --type=NodePort --port=80
    --target-port=8001 --dry-run -o yaml > nginx-nodeport.
    yaml
    ```

    The nginx-nodeport.yaml file will look like this:

    ```
    apiVersion: v1
    kind: Service
    metadata:
      labels:
        app: nginx
      name: nginx
    spec:
      ports:
      - port: 80
        protocol: TCP
        targetPort: 8001
      selector:
        app: nginx
      type: NodePort
    ```

2. Apply the generated file using the next command:

    ```
    kubectl create -f nginx-pod.yaml
    ```

Now that you have learned how to create a NodePort service for your application in a pod, it's time to learn how to use LoadBalancer services in the next section.

Exposing your pods using a LoadBalancer service and YAML files

To create a `LoadBalancer` service to expose your application inside a pod, follow these steps:

1. For `LoadBalancer`, use the next command:

   ```
   $ kubectl expose pod/nginx --type=LoadBalancer --port=80
   -target-port=8001 --dry-run -o yaml > nginx-lb.yaml
   ```

   The generated `nginx-lb.yaml` file will look like this:

   ```
   apiVersion: v1
   kind: Service
   metadata:
     labels:
       app: nginx
     name: nginx
   spec:
     ports:
     - port: 80
       protocol: TCP
       targetPort: 8001
     selector:
       app: nginx
     type: LoadBalancer
   ```

2. Apply the generated file using the next command:

   ```
   $ kubectl create -f nginx-pod.yaml
   ```

You have learned how to create a `LoadBalancer` service. With this, we have covered all the basic services in Kubernetes. Now, we are ready to learn how to create stateful applications. Let's move on to the next section to add persistence to your applications.

Adding persistence to your applications

Now, it is time to add storage to your applications; we are going to use the storage classes installed with Longhorn to provide persistence to your applications. In this section, we are going to explore two examples using persistent volumes. In this part of the book, we are going to discuss the persistent volumes and the process of creating storage for a Pod. But first, we need a persistent volume claim definition to provision this storage.

Creating an NGINX pod with a storage volume

To create your NGINX application using a storage volume that uses the Longhorn storage class, follow these steps:

1. Create pvc.yaml:

   ```
   apiVersion: v1
   kind: PersistentVolumeClaim
   metadata:
     name: longhorn-volv-pvc
   spec:
     accessModes:
       - ReadWriteMany
     storageClassName: longhorn
     resources:
       requests:
         storage: 2Gi
   ```

2. Apply the pvc.yaml YAML file:

   ```
   $ kubectl create -f pvc.yaml
   ```

   Now, it's time to create a pod using this PVC that uses the Longhorn storage class. To do this, follow these steps:

3. Create and apply the pod.yaml file to create a pod using the previously created PVC:

   ```
   apiVersion: v1
   kind: Pod
   metadata:
     name: volume-test
     namespace: default
   spec:
     containers:
     - name: volume-test
       image: nginx:stable-alpine
       imagePullPolicy: IfNotPresent
       volumeMounts:
       - name: volv
         mountPath: /data
   ```

```
    ports:
    - containerPort: 80
  volumes:
  - name: volv
    persistentVolumeClaim:
      claimName: longhorn-volv-pvc
```

This example has created a pod using a persistent volume, with the Longhorn storage class. Let's continue with a second example that shows a database using a storage volume.

Creating the database using a persistent volume

Now, is time to use a persistent volume for a database; in this example, you are going to learn how to create a Redis database with a persistent volume. So, let's get started with the following steps:

1. Create the redis.yaml file to create a pod that uses the previous longhorn-volv-pvc PVC:

```
apiVersion: v1
kind: Pod
metadata:
  name: redis
spec:
  containers:
  - name: redis
    image: redis
    volumeMounts:
    - name: redis-storage
      mountPath: /data/redis
  volumes:
  - name: redis-storage
    persistentVolumeClaim:
      claimName: longhorn-volv-pvc
```

2. Apply the pod.yaml YAML file to create the pod:

```
$ kubectl create -f pod.yaml
```

3. Check and Apply the `pod.yaml` YAML file to create the pod:

    ```
    $ kubectl create -f pod.yaml
    ```

4. Apply the `pod.yaml` YAML file to create the pod:

    ```
    $ kubectl create -f pod.yaml
    ```

Troubleshooting Your Deployments

Remember that you can use the `kubectl logs` command to troubleshoot your deployments. For more information, you can check the next link: `https://kubernetes.io/docs/tasks/debug-application-cluster/debug-running-pod/`.

Now, your Redis database is running and using a persistent volume to prevent the loss of data. In the last section, we are going to explore how to install a simple Kubernetes dashboard to manage your cluster using a UI.

Deploying a Kubernetes dashboard

Now, it's time to install a Kubernetes dashboard. The next steps are based on the official K3s documentation. To start installing the dashboard, follow these steps:

1. Install the dashboard using the following commands:

    ```
    $ GITHUB_URL=https://github.com/kubernetes/dashboard/
    releases
    $ VERSION_KUBE_DASHBOARD=$(curl -w '%{url_effective}'
    -I -L -s -S ${GITHUB_URL}/latest -o /dev/null | sed -e
    's|.*/||')
    $ sudo k3s kubectl create -f https://raw.
    githubusercontent.com/kubernetes/dashboard/${VERSION_
    KUBE_DASHBOARD}/aio/deploy/recommended.yaml
    ```

    This is going to install the dashboard, but you need to configure how to access this dashboard.

2. Create the `dashboard-admin-user.yaml` file to create a service account that provides access to your dashboard. The content of this file will be as follows:

    ```
    apiVersion: v1
    kind: ServiceAccount
    metadata:
      name: admin-user
      namespace: kubernetes-dashboard
    ```

3. Now create the file `dashboard-admin-user-role.yaml`. The content of this file will be the next:

    ```
    apiVersion: rbac.authorization.k8s.io/v1
    kind: ClusterRoleBinding
    metadata:
      name: admin-user
    roleRef:
      apiGroup: rbac.authorization.k8s.io
      kind: ClusterRole
      name: cluster-admin
    subjects:
    - kind: ServiceAccount
      name: admin-user
      namespace: kubernetes-dashboard
    ```

4. Now, apply the YAML files with the following command:

    ```
    $ kubectl create -f dashboard-admin-user.yml -f
    dashboard-admin-user-role.yml
    ```

5. Get the token inside the service account that will be used to access the dashboard:

    ```
    $ kubectl -n kubernetes-dashboard describe secret admin-
    user-token | grep '^token'
    ```

    Copy the token content only.

6. Use `kubectl proxy` to expose the Kubernetes API in your localhost, using the following command:

    ```
    $ sudo kubectl proxy
    ```

7. Access your browser with the following URL:

    ```
    http://localhost:8001/api/v1/namespaces/kubernetes-dashboard/
    services/https:kubernetes-dashboard:/proxy/
    ```

Sign in with the admin user bearer token that you got. Choose the **Token** option and enter the token. You will see a screen like this:

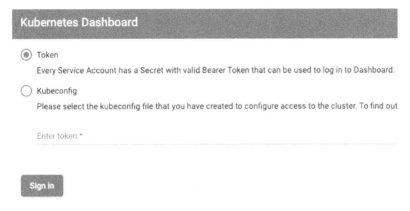

Figure 5.3 – Kubernetes Dashboard sign-in screen

After clicking on the **Sign In** button, you will see the dashboard. Explore the different menus to see the state of your objects, or click on the plus icon at the lower-right corner to create objects using the YAML files:

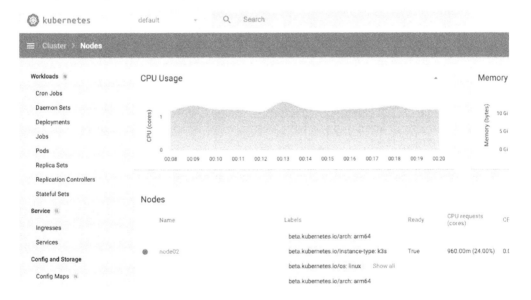

Figure 5.4 – Kubernetes Dashboard showing CPU and memory usage

We have now completed all the necessary concepts, giving you a quick introduction to how to use basic objects in Kubernetes with K3s.

Summary

In this chapter, we learned how to set up a K3s cluster with Raspberry Pi devices for our homelab. We also covered how to use basic Kubernetes objects to deploy an application. We deployed sample applications in an imperative way using the `kubectl` command. We also deployed sample applications using YAML files too. At the end of the chapter, we covered how to install a Kubernetes dashboard to manage your cluster. In the next chapter, we are going to continue adding more pieces to this deployment; we are going to use ingress controllers to deploy applications at the edge.

Questions

Here are a few questions to validate your new knowledge:

1. What are the basic Kubernetes objects that I need to create an application?
2. How can I install a K3s cluster for my homelab?
3. How can I use `kubectl` to create my applications?
4. How can I use YAML files to create my applications?
5. How can I use persistent volumes?
6. How can I troubleshoot my applications?

Further reading

You can refer to the following references for more information on the topics covered in this chapter:

- K3s installation options to add custom parameters to your config files: https://rancher.com/docs/k3s/latest/en/installation/install-options
- Longhorn official page: https://longhorn.io
- MetalLB official page: https://metallb.universe.tf
- Official Kubernetes documentation: https://kubernetes.io/docs
- Kubernetes Dashboard installation guide: https://rancher.com/docs/k3s/latest/en/installation/kube-dashboard
- Kubernetes Dashboard installation using Helm: https://artifacthub.io/packages/helm/k8s-dashboard/kubernetes-dashboard

Part 2:
Cloud Native Applications at the Edge

Here you will learn how to deploy your applications at the edge using GitOps, service meshes, serverless and event-driven architectures, and different types of databases.

This part of the book comprises the following chapters:

- *Chapter 6, Exposing Your Applications Using Ingress Controllers and Certificates*
- *Chapter 7, GitOps with Flux for Edge Applications*
- *Chapter 8, Observability and Traffic Splitting Using Linkerd*
- *Chapter 9, Edge Serverless and Event-Driven Architectures with Knative and Cloud Events*
- *Chapter 10, SQL and NoSQL Databases at the Edge*

6

Exposing Your Applications Using Ingress Controllers and Certificates

Ingress controllers fulfill traffic rules defined by an ingress object and are needed to expose traffic to APIs or microservices that your system uses. Ingress controllers are implemented in Kubernetes clusters. As an option to expose your deployments outside the cluster, instead of using dedicated load balancers for each deployment, the ingress controller shares a single load balancer for your deployments. By default, Kubernetes uses ClusterIP services to access deployments in the internal cluster network. Creating applications for edge computing involves configuring ingress controllers with lightweight solutions in mind: K3s and its default ingress controller, **Traefik** v1 using other ingress controllers such as **NGINX** or **Contour**, and security encryption for service communication.

In this chapter, we're going to cover the following main topics:

- Understanding ingress controllers and ingresses
- Installing Helm for ingress controller installations
- Installing and configuring cert-manager
- Using Traefik to expose your applications
- Using NGINX to expose your applications
- Troubleshooting your ingress controllers
- Pros and cons of Traefik, NGINX, and Contour
- Tips and best practices for ingress controllers

Technical requirements

Before starting, you need the following to run the examples in this chapter:

- A Raspberry Pi cluster with K3s installed
- kubectl configured to access your cluster
- Helm installed and configured

> **Note**
>
> If you don't want to use Traefik and you want to omit the default installation of this ingress controller in your cluster, add the `--no-deploy traefik --disable traefik` flags when you are installing your master node. For other details of installing your K3s cluster, refer to *Chapter 3, K3S Advanced Configurations and Management*, or visit `https://rancher.com/docs/k3s/latest/en/installation/install-options/server-config/`. Remember to install a bare metal load balancer such as MetalLB, which is necessary to generate a load balancer service, which is needed to install ingress controllers.

With these requirements, you are going to experiment with exposing your applications in different ways.

For more detail and code snippets, check out this resource on GitHub: `https://github.com/PacktPublishing/Edge-Computing-Systems-with-Kubernetes/tree/main/ch6`

Understanding ingress controllers

Kubernetes uses ingress controllers to expose your deployments outside the cluster. An ingress controller is the adaptation of a proxy to expose your applications, and Ingress is the Kubernetes object that uses this adaptation. An ingress controller works as a reverse proxy like **NGINX** to expose your application using HTTP/HTTPS protocols to a load balancer. This load balancer is the endpoint to expose your application outside the cluster. It's in charge of receiving and controlling traffic for your application. The benefit of this is that you can share this load balancer, to expose as many applications as you want, but using all the features that your ingress controller provides. There are different ingress controller implementations, such as NGINX, Traefik, Emissary, and Envoy.

Taking as a reference *Figure 6.1*, to expose your application, you must create a **ClusterIP** service that creates an internal DNS name for your Deployment or Pod. This service automatically forwards the traffic across the different replicas of your service, which perform load balancing. An Ingress uses the LoadBalancer service that your ingress controller provisioned when you installed it. This LoadBalancer has a public IP address if the cluster is not private. This IP receives traffic outside the cluster, then forwards this traffic to the ClusterIP service that your application is using. Internally, the Ingress object uses configuration files to act as a reverse proxy. For example, if you are using NGINX, the ingress object is going to use configurations that are used in a regular NGINX configuration file.

In the context of Kubernetes, an ingress object tries to match the associated ClusterIP service of your application, using labels. This is how an ingress works internally. You can see an ingress as the common virtual hosts feature that NGINX and Apache provide for websites.

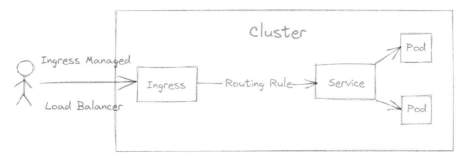

Figure 6.1 – Ingress in Kubernetes

Based on the official documentation of Kubernetes, a basic YAML file to create an ingress controller should look like this:

```
apiVersion: networking.k8s.io/v1
kind: Ingress
metadata:
  name: minimal-ingress
  annotations:
    kubernetes.io/ingress.class: nginx
    nginx.ingress.kubernetes.io/rewrite-target: /
spec:
  rules:
  - http:
      paths:
      - path: /testpath
        pathType: Prefix
        backend:
          service:
            name: test
            port:
              number: 80
```

The most important sections for ingresses are the annotations and spec sections. The annotations will define the ingress controller to use, in this case, NGINX. This section could include additional configurations for the ingress, such as rewriting the URL or activating features such as authentication, and so on. This example defines the /testpath route to access your application. Now you have to apply the YAML file with kubectl. For example, if this file is called minimal-ingress.yaml, you have to run the following command to create the minimal-ingress controller:

```
$ kubectl apply -f minimal-ingress.yaml
```

And that's the way that ingress controllers and ingresses work. Now let's install Helm to install an ingress controller in the next section.

Installing Helm for ingress controller installations

Before we start using an ingress controller, we need to install Helm. Helm is a package manager for Kubernetes, which you can use to install software. Helm uses Helm Charts, which contain the definitions to install and configure your deployments.

To install Helm, follow the given steps:

1. Download Helm with the next command:

   ```
   $ curl -fsSL -o get_helm.sh https://raw.
   githubusercontent.com/helm/helm/main/scripts/get-helm-3
   ```

2. Change permissions and launch the installer by executing the following lines:

   ```
   $ chmod 700 get_helm.sh
   $ ./get_helm.sh
   ```

Now you have Helm installed, let's move on to install the NGINX ingress controller in the next section.

Installing cert-manager

cert-manager is software that you want to install if you want to add certificates or certificate issues as a resource type in Kubernetes. These certificates can be used by applications, but in this specific case, we are going to use cert-manager to add encrypted traffic to your app, using the HTTPS protocol.

To install cert-manager, we are going to use Helm. To install Helm, you have to follow the given steps:

1. Add the Jetstack repo, which contains cert-manager:

   ```
   $ helm repo add jetstack https://charts.jetstack.io
   ```

2. Update your local Helm Chart repository cache. To do this, execute the following command:

```
$ helm repo update
```

3. Now install cert-manager using Helm:

```
$ helm install \
cert-manager jetstack/cert-manager \
--namespace cert-manager \
--create-namespace \
--version v1.5.4 \
--set prometheus.enabled=false \
--set webhook.timeoutSeconds=4 \
--set prometheus.enabled=false \
--set installCRDs=true
```

This is going to install cert-manager in the cert-manager namespace, with version 1.5.4. This cert-manager installation doesn't include Prometheus but includes cert-manager **Customer Resource Definitions** (**CRDs**) and configures timeout parameters for webhook validations when generating a certificate.

4. Create a self-signed issuer in cert-manager, to create certificates for your local domains. To do this, create the self-signed-issuer.yaml file with the following content:

```
apiVersion: cert-manager.io/v1
kind: ClusterIssuer
metadata:
  name: self-signed-issuer
spec:
  selfSigned: {}
```

5. Now create an issuer that uses Let's Encrypt to create a certificate that could be used for public domains. To do this, create the letsencrypt-staging.yaml file with the following content:

```
apiVersion: cert-manager.io/v1
kind: Issuer
metadata:
  name: letsencrypt-staging
spec:
  acme:
    server: https://acme-staging-v02.api.letsencrypt.org/
directory
```

```
      email: user@example.com
      privateKeySecretRef:
        name: letsencrypt-staging
      solvers:
      - http01:
          ingress:
            class:   nginx
```

This file is assuming, as an example, that you will use this issuer in a staging environment, but you can customize this file as you want.

> **Important Note**
> Be aware that `ClusterIssuer` is cluster scoped and `Issuer` is namespace scoped.

6. Now apply these files to create the self-signed issuer for a staging environment, using the following command:

    ```
    $ kubectl apply -f self-signed-issuer.yaml -f
    letsencrypt-staging.yaml
    ```

Now you have cert-manager installed and ready to use. You can also create basic issuers for your applications. This part will be crucial to configure certificates for your domains if necessary. So now, let's move towards installing our first ingress controller, NGINX.

NGINX ingress installation

NGINX is the most widely used ingress controller on Kubernetes. It has all the necessary features that you need for basic and complex configurations to expose your application. It has all the experience and support of the community behind NGINX. It's stable and you can still use it for devices using ARM processors.

To install the NGINX ingress controller, follow the given steps:

1. Create a namespace to install the NGINX ingress controller:

    ```
    $ kubectl create ns nginx-ingress
    ```

2. Add the repository that contains the Helm Chart of the NGINX ingress controller and update the repository of charts that Helm is going to use:

    ```
    $ helm repo add ingress-nginx https://kubernetes.github.
    io/ingress-nginx
    $ helm repo update
    ```

3. Install the NGINX ingress controller with the following command:

```
$ helm install nginx-ingress ingress-nginx/ingress-nginx
-n nginx-ingress
```

This will output that the installation was successful.

4. To check whether the `nginx-ingress` controller was installed, run the following command:

```
$ helm list -n nginx-ingress
```

5. After installing the `nginx-ingress` controller, K3s will provide a load balancer. In this case, we assume that we are using MetalLB. To obtain the load balancer IP address provisioned by your ingress controller, run the following command:

```
$ IP_LOADBALANCER=$(kubectl get svc nginx-ingress-
ingress-nginx-controller --output jsonpath='{.status.
loadBalancer.ingress[0].ip}' -n nginx-ingress)
```

Here, the `IP_LOADBALANCER` variable contains the IP of the load balancer created by the ingress controller, which is the endpoint for your applications. You can check the value by running the following command:

```
$ echo $IP_LOADBALANCER
```

Using as a reference the installation of the K3s cluster in *Chapter 5*, *K3s Homelab for Edge Computing Experiments*, you will see an IP like this: `192.168.0.240`.

6. You can use that IP to create a DNS record to point the ingress to a domain, or to access your service using a path. Let's say that, for example, the returned IP is `192.168.0.241`. You can access your service access in your browser with the URL `http://192.168.0.240`. Another option is to use a path to access your application; for example, the URL would be like this: `http://192.168.0.240/myapp`.

7. Finally, if you want to uninstall `nginx-ingress`, run the following command:

```
$ helm uninstall nginx-ingress -n nginx-ingress
```

Now that you have installed the NGINX ingress controller, let's move on to study a basic example using this ingress controller.

Using NGINX to expose your applications

It's time to start using NGINX as your ingress controller. We are going to expose your first application using NGINX. To begin, let's deploy a simple application. To do this, follow the given steps:

1. Create a simple deployment using `nginx` image with the following command:

    ```
    $ kubectl create deploy myapp --image=nginx
    ```

2. Create a ClusterIP service for the `myapp` deployment:

    ```
    $ kubectl expose deploy myapp --type=ClusterIP --port=80
    ```

3. Create an Ingress using the domain `192.168.0.240.nip.io`. In this example, we are assuming that the endpoint for the ingress is `192.168.0.240`. This is the same IP as the load balancer created by the ingress controller. When you access your browser, the page `https://192.168.0.241.nip.io` is going to show the NGINX **myapp** Deployment, which you have already created. `nip.io` is a wildcard DNS for any IP address, so with this, you can get a free kind of domain to play with your ingress definitions. Let's move on to create this ingress by creating the `myapp-ingress.yaml` file:

    ```
    apiVersion: networking.k8s.io/v1
    kind: Ingress
    metadata:
      name: myapp-ingress-tls-nginx
      annotations:
        kubernetes.io/ingress.class: "nginx"
        cert-manager.io/cluster-issuer: self-signed-issuer
    spec:
      tls:
      - hosts:
          - 192.168.0.241.nip.io
        secretName: myapp-tls-nginx
      rules:
      - host: 192.168.0.241.nip.io
        http:
          paths:
          - path: /
            pathType: Prefix
            backend:
              service:
    ```

```
    name: myapp
    port:
      number: 80
```

You can customize this file as you want. If you don't want HTTPS for your application, omit the TLS section and the annotation `cert-manager.io/cluster-issuer`. If you are using a public domain, use the following annotation:

```
cert-manager.io/cluster-issuer: letsencrypt-staging
```

4. If you are using a local domain, for example, `myapp-test-nginx.test`, you have to modify the `/etc/hosts` file and add a line like this:

```
192.168.0.241 myapp-test-nginx.test
```

This is necessary to resolve the local domain for your app. Also, remember to modify `tls.hosts` and `rules.hosts` in the file in order to use a domain such as `myapp-test-nginx.test`. So, the second option will be like this:

```
apiVersion: networking.k8s.io/v1
kind: Ingress
metadata:
  name: myapp-ingress-tls-nginx
  annotations:
    kubernetes.io/ingress.class: "nginx"
    cert-manager.io/cluster-issuer: self-signed-issuer
spec:
  tls:
  - hosts:
      - myapp-test-nginx.test
    secretName: myapp-tls-nginx
  rules:
  - host: myapp-test-nginx.test
    http:
      paths:
      - path: /
        pathType: Prefix
        backend:
          service:
```

```
        name: myapp
        port:
            number: 80
```

5. Create the ingress using the previous file using the following command:

    ```
    $ kubectl apply -f myapp-ingress-tls-nginx.yaml
    ```

6. Access the `myapp` deployment by using the URL `https://192.168.0.241.nip.io` or `https://myapp-test-nginx.test` in your browser page.

> **Note**
>
> Because this is a self-signed certificate, you have to accept the security exception in your browser.

   Or, use `curl` to access the page with the following command:

    ```
    $ curl -k https://192.168.0.240.nip.io
    or
    $ curl -k https://myapp-test-nginx.test
    ```

   If you don't want to use HTTPS, you can access the page with the URL `http://192.168.0.241.nip.io` or `https://myapp-test-nginx.test` in your browser or by using the `curl` command.

7. If you want to delete the ingress, run the following command:

    ```
    $ kubectl delete -f myapp-ingress.yaml
    ```

> **Note**
>
> When you delete the ingress, and you are using a self-signed issuer, the secret used for *Let's Encrypt* will not be deleted. You have to delete it manually using the `kubectl` command. For example, you can run the following command to delete the secret from the previously created ingress: `kubectl delete secrets myapp-tls-nginx`.

Now you have learned how to use NGINX. Next, it's time to learn how to use Traefik to expose your applications in the next section.

Using Traefik to expose your applications

Traefik is the ingress controller included by default in K3s. It uses the same configurations as NGINX as shown in the previous example in the `myapp-ingress.yaml` file. Let's assume that you already have created the `myapp` Deployment from the previous section. So, let's get started with Traefik by following the given steps:

1. To find the load balancer IP address created by Traefik, run the following command:

   ```
   $ IP_LOADBALANCER=$(kubectl get svc traefik --output
   jsonpath='{.status.loadBalancer.ingress[0].ip}' -n kube-
   system)
   ```

   Run the following command to see the current IP address assigned to the load balancer that the Traefik installation provisioned. This will be used to create an entry in the `/etc/hosts` file:

   ```
   $ echo $IP_LOADBALANCER
   ```

   Let's say that returns `192.168.0.240`. You have to add the next line to the `/etc/hosts` file:

   ```
   192.168.0.240 myapp-test-traefik.test
   ```

   Now you are ready to create the Ingress object.

2. To expose `myapp` using `nip` and TLS, create the `myapp-ingress-tls-traefik.yaml` file with the following content:

   ```yaml
   apiVersion: networking.k8s.io/v1
   kind: Ingress
   metadata:
     name: myapp-ingress-traefik
     annotations:
       kubernetes.io/ingress.class: "traefik"
       cert-manager.io/cluster-issuer: self-signed-issuer
       traefik.ingress.kubernetes.io/router.tls: "true"
   spec:
     tls:
     - hosts:
         - myapp-test-traefik.test
       secretName: myapp-tls-traefik
     rules:
     - host: myapp-test-traefik.test
       http:
         paths:
   ```

```
    - path: /
      pathType: Prefix
      backend:
        service:
          name: myapp
          port:
            number: 80
```

3. Apply the file with the following command:

```
$ kubectl apply -f myapp-ingress-tls-traefik.yaml
```

4. (*Optional*) If you want to use the nip.io service, the YAML file will look like this:

```
apiVersion: networking.k8s.io/v1
kind: Ingress
metadata:
  name: myapp-ingress-traefik
  annotations:
    kubernetes.io/ingress.class: "traefik"
    cert-manager.io/cluster-issuer: self-signed-issuer
    traefik.ingress.kubernetes.io/router.tls: "true"
spec:
  tls:
  - hosts:
      - myapp-test-traefik.test
    secretName: myapp-tls-traefik
  rules:
  - host: myapp-test-traefik.test
    http:
      paths:
      - path: /
        pathType: Prefix
        backend:
          service:
            name: myapp
            port:
              number: 80
```

5. (*Optional*) Apply the file with the following command:

```
$ kubectl apply -f myapp-ingress-tls-traefik.yaml
```

Now you have configured and used Traefik as your load balancer.

Remember that if you didn't use the `-disable traefik` parameter, Traefik will be installed in your K3s cluster. Now, it's time to use Contour. So, let's move on to the next section.

Contour ingress controller installation and use

Contour is an Envoy-based ingress controller. The advantage of using Envoy is that it's fast and includes some powerful features that are found in service meshes, such as rate limits, advanced routing, metrics, and so on. If speed is key in your project, Contour will be the best solution in most cases. Contour is a lightweight solution and is optimized to run quickly. This makes Contour a good choice for edge computing. Now let's move on to start using Contour.

To install Contour, follow the next steps:

1. Install Contour using the quickstart configuration it provides:

```
$ kubectl apply -f https://projectcontour.io/quickstart/
contour.yaml
```

2. If you want to use `nip.io`, you have to first find the IP of the Contour load balancer and create an entry in the `/etc/hosts` file from your machine. To find the IP of Contour, run the following command:

```
$ IP_LOADBALANCER=$(kubectl get svc envoy--output
jsonpath='{.status.loadBalancer.ingress[0].ip}' -n
projectcontour)
```

If you run the following command, it will show the load balancer IP that the Contour ingress controller installation provisioned, which will be used to create an entry in the `/etc/hosts` file:

```
$ echo $IP_LOADBALANCER
```

This will show the load balancer IP that the Contour installation created. This will be used to create an entry in the /etc/hosts file.

Let's say that returns 192.168.0.242. You have to add the next line to the /etc/hosts file:

```
192.168.0.242 myapp-test-contour.test
```

Now you are ready to create the Ingress object.

3. Create a file with a basic configuration for contour. Let's call this file myapp-ingress-tls-contour.yaml. This file will have the following content:

```
apiVersion: networking.k8s.io/v1
kind: Ingress
metadata:
  name: myapp-ingress-tls-contour
  annotations:
    kubernetes.io/ingress.class: "contour"
    cert-manager.io/cluster-issuer: self-signed-issuer
spec:
  tls:
  - hosts:
      - myapp-test-contour.test
    secretName: myapp-tls-contour
  rules:
  - host: myapp-test-contour.test
    http:
      paths:
      - path: /
        pathType: Prefix
        backend:
          service:
            name: myapp
            port:
              number: 80
```

4. Apply the YAML file with the following command:

```
$ kubectl apply -f myapp-ingress-tls-contour
```

Now we know how to use Contour using the Ingress object in Kubernetes. So, let's see how to use Contour using its own objects in Kubernetes in the next section.

Using Contour with HTTPProxy and cert-manager

Contour can be used in the same way as the NGINX ingress controller, but you can also use the HTTPProxy object that Contour provides. The same example, myapp-ingress-tls-contour, can be created using Contour objects. Let's see the equivalent for the Contour ingress controller. First, let's create the certificate with the cert-manager object. Let's call the file myapp-tls-contour. yaml. It will look like this:

```
apiVersion: cert-manager.io/v1
kind: Certificate
metadata:
  name: myapp-tls-contour
spec:
  commonName: myapp-test-contour.test
  dnsNames:
  - myapp-test-contour.test
  issuerRef:
    name: self-signed-issuer
    kind: ClusterIssuer
  secretName: httpbinproxy
```

The myapp-tls-contour.yaml file definition creates the certificate to be used by the HTTPProxy object. Let's create the myapp-ingress-http-proxy-tls-contour.yaml file with the equivalent configuration of the myapp-ingress-tls-contour.yaml file, but now using the HTTPProxy object and the previously generated certificate. This will look like this:

```
myapp-ingress-http-proxy-tls-contour.yaml
apiVersion: projectcontour.io/v1
kind: HTTPProxy
metadata:
  name: myapp-ingress-http-proxy-tls-contour
spec:
  virtualhost:
    fqdn: myapp-test-contour.test
    tls:
      secretName: myapp-tls-contour
```

```
rateLimitPolicy:
  local:
    requests: 3
    unit: minute
    burst: 1
routes:
- services:
  - name: myapp
    port: 80
```

Notice that this object sets a rate limit of 3 requests per minute with an additional request or soft limit to have, in total, 4 requests per minute. If the limit is exceeded, Contour will block the request. You can access the site with the following command:

```
$ curl -k https://myapp-test-contour.test
```

The -k parameter omits the validation of the self-signed certificate created by cert-manager.

As you can see, Contour can use the Kubernetes ingress object, and you can add more features as rate limits using the objects provided by Contour. Now, it's time to troubleshoot your ingress controllers or ingress definitions. Let's move on to the next section.

Troubleshooting your ingress controllers

These are some useful commands that you can use to troubleshoot your ingress controllers:

1. To check the NGINX ingress controller logs, run the following command:

   ```
   $ kubectl logs -f deploy/nginx-ingress-ingress-nginx-
   controller -n nginx-ingress
   ```

   This will show the logs when an ingress uses NGINX as the ingress controller.

2. To check Traefik ingress controller logs, run the following command:

   ```
   $ kubectl logs -f deploy/traefik -n kube-system
   ```

3. To check Contour ingress controller logs, run the following command:

```
$ kubectl logs -f deploy/contour -n projectcontour
```

These commands are useful for checking what is happening inside your ingress controller deployments. Now, here are some useful commands to check that your ingress definition is working properly:

```
$ kubectl get svc
$ kubectl get ingresses
```

If you want to use the ingress controller's own objects, such as HTTPProxy and so on, run the following command:

```
$ kubectl get OBJECT_NAME
```

Here, OBJECT_NAME should be, for example, HTTPProxy, Certificate, and so on. This depends on what object you want to check. For a full list of these objects, you can check the official documentation for NGINX, Traefik, and Contour.

Now you have learned about troubleshooting your ingress controller deployment and your ingress definitions, let's explore the pros and cons of the ingress controllers that we have used in this chapter.

Pros and cons of Traefik, NGINX, and Contour

All the ingress controllers have the basic features to expose your application, that is, they are compatible with the Ingress object in Kubernetes. So, let's explore the pros and cons of each Ingress controller. Let's get started with this quick comparison:

- NGINX Ingress is an ingress controller that uses NGINX to expose applications in your cluster.

    - Pros: It is the most widely used ingress controller for Kubernetes. It has a lot of documentation. Developer and community support is widely available. The community behind it is bigger than Traefik and Contour.

    - Cons: It can be slow compared to Envoy-based ingress controllers such as **Emissary**, **Gloo**, and Contour.

- Traefik is an ingress controller created by Traefik Labs. It has a lot of features, which can be used as plugins. It can be used to visualize your applications on a dashboard.

    - Pros: It has a dashboard and a lot of documentation. It also has some service mesh capabilities.

    - Cons: It can be slow when compared against NGINX and Contour. The documentation is not focused on Kubernetes, and can be difficult to understand.

- Contour is an ingress controller based on Envoy, a tool owned by VMware. It's used in Tanzu, a platform for managing Kubernetes. This means that a big company supports Contour.

  - Pros: It's fast because of its architecture and the language used for its binary, which is C. It has enough features to expose your application. It can be used as a service mesh. Big projects such as Istio use Envoy as their default ingress controller. Contour has support for ARM devices.

  - Cons: Contour is not mature and has missing features. It has fewer features compared with NGINX and Traefik.

The use of ingress controllers could be focused on exposing your application. Depending on the feature you need, you can choose the previous ingress controllers. If you want to use a stable ingress controller, choose NGINX. If you are looking for auto-discovery features or a dashboard to visualize your endpoints, maybe you could use Traefik. And if you are looking for speed or a customizable ingress controller, choose Contour, or maybe you can create your own solution using Envoy.

Tips and best practices for ingress controllers

These are some ideas that you can explore when using ingress controllers:

- **Use routing features**: Each of these ingress controllers has different ways to implement routing to expose your application. Read the official documentation of these ingress controllers to understand which has your desired features.

- **Create a proof of concept** (POC) to evaluate which ingress controller is best for your use case.

- **Install Traefik 2.0**: If you like Traefik, maybe you can install Traefik 2.0. K3s includes Traefik version 1.0, which only has the necessary features to expose your application. But if you need more advanced reverse proxy features for your applications, you can install Traefik 2.0, which includes a dashboard and other features that you may want to use.

- **Introduce rate limits**: Implement rate limits to your applications. This is a nice feature when you want to prevent spikes or denial-of-service attacks.

- **Implement TLS**: This is a common use case. It's recommended to encrypt your traffic to prevent a hacker from stealing your information. It's important to provide additional security for your applications.

- **Install basic authentication**: This is the most basic kind of security for your endpoints. With this, you can set a user and password to access your applications.

- **Secure access with JSON Web Tokens** (JWTs): This is a nice feature to get more control and use tokens to access your endpoints. It is a better and more secure option than using a basic authentication method.

Now you have other ideas to implement when you are using an ingress controller and creating ingress definitions to expose your applications. Now it is time for a quick summary of this chapter.

Summary

In this chapter, we learned how to use different ingress controllers, such as NGINX, Traefik, and Contour. These ingress controllers are the most used ones, starting with NGINX, then Traefik, and finally Contour, which is based on Envoy. This chapter showed you how to use NGINX, Traefik, and Contour to solve common daily tasks in real production environments. The examples covered the use of TLS, routes, and some basic limit rates to access your applications. This chapter covered the last topic necessary to start with practical applications of all these technologies in the next chapter.

Questions

Here are a few questions to validate your new knowledge:

- What is an ingress controller?

- When can you use an ingress controller?

- How can you create an ingress definition to expose your applications?

- How can you create your ingress definition for NGINX, Traefik, or Contour?

- How can you troubleshoot your ingress controllers and ingress definitions?

- How can you use MetalLB with your ingress controllers?

Further reading

You can refer to the following references for more information on the topics covered in this chapter:

- Kubernetes Ingress documentation: `https://kubernetes.io/docs/concepts/services-networking/ingress`

- Install cert-manager with Helm: `https://cert-manager.io/docs/installation/helm`

- Generating certificates for an ingress with cert-manager: `https://cert-manager.io/docs/tutorials/acme/nginx-ingress`

- Kubernetes ingress controller official documentation: `https://kubernetes.io/docs/concepts/services-networking/ingress`

- Installing NGINX ingress controller: `https://kubernetes.github.io/ingress-nginx/deploy/#using-helm`

- Contour ingress controller getting started: `https://projectcontour.io/getting-started`

- Contour rate limits: `https://projectcontour.io/docs/v1.15.2/config/rate-limiting`

- Create a Kubernetes TLS Ingress from scratch in Minikube: `https://www.youtube.com/watch?v=7K0gAYmWWho`

- Traefik and Kubernetes: `https://doc.traefik.io/traefik/v1.7/configuration/backends/kubernetes`

- JWT generator: `https://jwt.io`

7

GitOps with Flux for Edge Applications

Previous chapters have already covered the basics of building your home lab using K3s. It's time to implement simple use cases that you can use in edge computing. This chapter covers how to implement GitOps for your applications using Flux in edge computing environments, starting with the basic theory of GitOps and the necessary tools to manage a Git repository for deployments. Then, we will look at how to install Flux to implement a basic GitOps workflow for a demo application. This chapter includes how to automate an application deployment using a **mono repository** (**monorepo**) configuration, the Helm operator, and the image updater feature of Flux. Finally, we will end the chapter with the installation of basic monitoring dashboards in Flux, essential troubleshooting commands for Flux, and how to uninstall Flux.

In this chapter, we're going to cover the following main topics:

- Implementing GitOps for edge computing
- Flux and its architecture
- Designing GitOps with Flux for edge applications
- Building your container image with GitHub Actions
- Installing and configuring Flux for GitOps
- Troubleshooting Flux installations
- Installing Flux monitoring dashboards
- Uninstalling Flux

Technical requirements

In this chapter, to implement GitOps using Flux, you will need the following:

- Three single node K3s clusters using a device with an **ARM** processor such as a Raspberry Pi.

- Previous experience with Git.

- GitHub repository and its token; you also need some basic experience using Git.

- Docker Hub account to push new image releases of your application.

With this, you are ready for this first use case to implement GitOps at the edge using Flux. So, let's get started.

For more detail and code snippets, check out this resource on GitHub: `https://github.com/PacktPublishing/Edge-Computing-Systems-with-Kubernetes/tree/main/ch7`

Implementing GitOps for edge computing

To start this topic, let's get started with the concept of GitOps. The `https://www.gitops.tech/` website states: *"GitOps is a way of implementing Continuous Deployment for cloud native applications. It focuses on a developer-centric experience when operating infrastructure, by using tools developers are already familiar with, including Git and Continuous Deployment tools."* This means that GitOps helps you with your **continuous deployment** (**CD**) in general. In software engineering, it is common to refer to continuous deployment and continuous delivery with the CD acronym.

Also, the GitLab page `https://about.gitlab.com/topics/gitops` mentions that GitOps contains the following basic components:

- **Infrastructure as code** (**IaC**): This refers to a declarative way to provide infrastructure or deployments for your applications.

- **Merge requests or pull requests** (**PRs**): A way to manage infrastructure or application code updates across multiple changes and collaborators.

- **Source code management** (**SCM**): Systems such as Git enable merge request- or pull request-based workflows and a mechanism to manage this, usually using a Git repository. In this way, a team can have an approval-and-review mechanism to apply changes.

- **Continuous integration and continuous delivery** (**CI/CD**): CI and CD include, by nature, all the processes of building, checking, and deploying applications and changes to those software applications. GitOps is used to automate CD for a cloud native application.

In this chapter, you will find tools that provide mechanisms for CI/CD automation pipelines. Let's pay attention to the following diagram that shows a summary of how GitOps works:

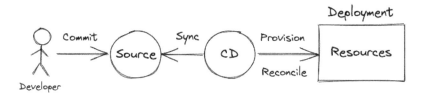

Figure 7.1 – GitOps

The basics of starting with GitOps are outlined here, as depicted in *Figure 7.1*:

1. **Commit code to the source**: Here, a developer makes changes and pushes their changes to a source or repository—for example, a Git repository hosted at GitHub.

2. **Synchronize changes**: A GitOps tool such as, for example, Flux periodically detects changes in the source.

3. **Provision or reconcile changes**: Once the GitOps tool detects changes, it aims to automate the process of updating a deployment based on the declarative configuration changes found. This could include processes to modify, such as changing configurations, updating the container image for an application using containers, and so on. Sometimes, if a resource or deployment doesn't exist, you have to provision resources or reconcile these, comparing changes. This means that a GitOps tool regularly works with declarative definitions to reflect the state of your infrastructure or application.

Finally, the user will see changes when accessing their application, and you can also add some additional processes such as notifications if a process was finalized, or an alert if something fails.

In general, this is how a GitOps process works and how a GitOps tool operates. In this chapter, we are going to focus on Flux as our GitOps tool, to implement GitOps processes to automate our applications' deployments and updates.

GitOps principles

There are some principles that you have to follow if you are using GitOps to automate your CD pipelines. Based on *Weaveworks*, these principles are as follows:

- **Declarative definitions**: You can often find these definitions using **YAML** files, but they could also be found in other formats such as **JSON**.

- **State of your applications versioned with Git**: GitOps tools use Git repositories to store changes and states for applications.

- **Approve changes that can be applied automatically to your resources**: Git repositories or services provide ways to automate a trigger tool when some changes or merges are detected.

- **Software agents listening to changes notifying or alerting**: GitOps tools also have daemons listening to changes in a ready-to-launch action, such as changing a repository with a new image tag in the case of applications using containers.

GitOps benefits

Now you know how a GitOps tool works and how GitOps processes help you to automate your CD pipeline, it's time to find out about the benefits of GitOps.

Based on *Weaveworks*, these are as follows:

- **Increased productivity**: An automated process reduces the execution time; in this case, more changes and updates made for your applications in less time.

- **Enhanced developer experience**: GitOps tools usually launch automated processes on your Git repository, and these will be launched automatically without knowing the internals of how it works—for example, for applications using Kubernetes, a developer doesn't have to know Kubernetes in some cases. However, this depends on how your application is structured.

- **Improved stability**: Logs of GitOps tools are included by default, which helps to meet some security and monitoring features.

- **Higher reliability**: GitOps tools give you the ability to implement rollback mechanisms, reducing downtime for your applications if a change has an impact on the operation of your system.

- **Consistency and standardization**: GitOps tools have structures to define your applications, and give you best practices for your applications' definitions, pipelines, or updates.

- **Stronger security guarantees**: GitOps tools have security features such as cryptography for secrets, and tracking and managing changes. This brings a way to secure your applications.

Now, let's move on to understand how GitOps works in a cloud native context.

GitOps, cloud native, and edge computing

As we know, **cloud native** refers to the use of applications using technologies such as containers, microservices, and CI/CD in the context of a **development-operations** (**DevOps**) culture. So, you can find an intersection for this concept when a GitOps tool is designed to run on cloud native environments—for example, Kubernetes clusters.

A GitOps tool can help you to automate the CD process for your Kubernetes applications. Tools such as Argo CD or Flux can help you to implement GitOps for your applications.

But in this chapter, we are going to focus more on applications that run in low-resource environments using ARM processors. In this case, Flux has support for ARM while Argo CD doesn't. This chapter focuses on implementing GitOps with Flux using ARM devices. So, let's get started with a brief introduction to Flux in the next section.

Flux and its architecture

The Flux website, `https://fluxcd.io`, says: "*Flux is a set of continuous and progressive delivery solutions for Kubernetes that are open and extensible.*" Flux gives you the ability to have your Kubernetes clusters in sync with the source that contains declarative definitions of your applications, commonly stored in Git repositories.

Flux also uses the Kubernetes API to manage its objects. It also uses its own GitOps Toolkit, which gives you the tools to build a CD system on top of Kubernetes. You can see how Flux works in the following diagram:

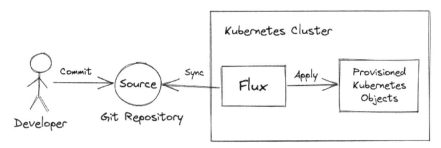

Figure 7.2 – Flux architecture

This diagram reflects a typical workflow for GitOps, starting with a commit and a GitOps tool that is constantly checking for changes in the application's definitions—in this case, YAML files. When Flux detects this change, it automatically provisions for the reconciliation of your applications, based on declarative definitions.

To bring essential functionalities to GitOps, Flux provides the following main features:

- Support for Git repositories of multiple providers
- Supported integrations for tools such as Kustomize and Helm
- Event-triggered and periodic reconciliation
- Integration with Kubernetes **role-based access control (RBAC)**
- Alerting external systems (webhook senders)

- External events handling (webhook receivers)

- Automated container image updates to Git (image scanning and patching)

As a GitOps tool for Kubernetes, Flux could be installed on ARM devices. In this way, Flux could be a good match for edge computing. But first, let's look at how Flux matches edge computing requirements.

Flux matches edge computing requirements for the following reasons:

- Has less complexity for GitOps compared with tools such as Argo CD, Tekton, and others

- Can be installed on ARM devices for low-resource environments

- Requires low resource consumption to operate

This is how Flux works and how it matches edge computing requirements. Now, let's see how we are going to organize our applications to implement GitOps for the edge using Flux in the next section.

Designing GitOps with Flux for edge applications

We are going to implement GitOps for edge computing with Flux, but first, we have to explain the whole workflow and the main parts of this implementation. For this, let's explore the following diagram, which explains the components and workflow of GitOps, implementing an image automation updater for your applications:

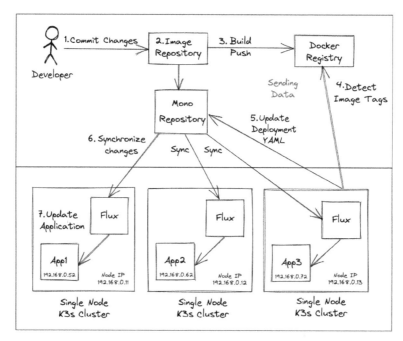

Figure 7.3 – Flux GitOps implementation using image updater feature

Our GitOps workflow implementation has the following steps:

1. A developer changes the application and submits changes with a PR to be merged into the main branch. You could make changes and push them directly to the main branch, but this is not a good practice since you may have submitted broken or unreviewed changes. In later examples of this chapter, we use GitHub to host our Git repository.

2. The repository has activated GitHub Actions and triggered a small pipeline just to build an image.

3. An image is built in the pipeline and tagged with a new version, then it is pushed to the public Docker registry. In most business scenarios, you have to use private repositories in the cloud or on-premises.

4. The image updater feature checks for new changes and tags for previous newly generated images that your application is going to use.

5. Once Flux detects the new image, it looks for files configured to be updated with the new image tag. Once Flux updates the files' definitions with the new tag, the changes are pushed to the repository.

6. Flux detects changes in the definition of files that were updated with the new image tag. Then, Flux triggers a reconciliation process to update your applications.

7. Objects in the Kubernetes cluster associated with the definition files are updated. Then, your application will run with the new image.

To implement the GitOps workflow just described, we are assuming the next networking and GitHub configurations:

*   Single node K3s clusters using the `192.168.0.0/24` network so that they can access the same network. You can use a different private network such as the `172.16.0.0/16` or `10.0.0.0/8` networks, for example.

*   Each cluster is using MetalLB as the bare metal load balancer service, using different IP ranges for load balancing. Cluster 1 is using IP addresses in the range of `192.168.0.51-60`, cluster 2 is using `192.168.0.61-70`, and cluster 3 is using `192.168.0.71-80` to do some basic IP address distribution for this network. The first addresses are typically used by the default load balancer of Traefik, so this IP address could be different in your network. Take a look at *Chapter 5, K3s Homelab for Edge Computing Experiments*, to configure MetalLB using the same or similar IP ranges.

*   You have a GitHub account and a token to access or create repositories in your account. Here's what we're doing:

    *   We are using the `https://github.com/sergioarmgpl/fluxappdemo` GitHub repository, which contains a basic Helm chart to deploy in our clusters. You can find more details about the application in the repository link.

With this, we are ready to start implementing this scenario in the next section.

Creating a simple monorepo for GitOps

For our GitOps implementation, we are going to use a monorepo. We have chosen to do this to reduce the management of many repositories and centralize all work in a single repository. For this use case, we are going to organize our cluster configurations and applications' definitions in a single repository. Let's explore the following screenshot to understand how our new repository will be organized:

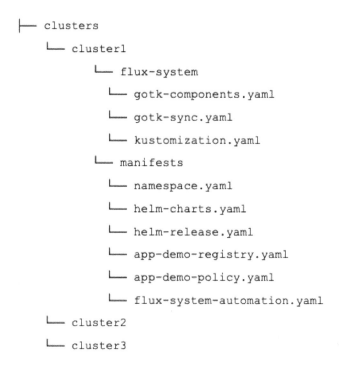

Figure 7.4 – Monorepo structure

Now let's describe what each directory and file does, as follows:

- clusters: This is the main directory that contains all the configuration of Flux and deployments in separated directories.

- cluster1-cluster3: Inside these folders, the definitions of Flux and your applications are organized. cluster1 will be the cluster in charge of updating YAML definitions for the application for all clusters. So, cluster2 and cluster3 don't need the image updater components in their installations.

- `flux-system`: Contains Flux definitions to deploy it. Includes the `gotk-components.yaml`, `gotk-sync.yaml`, and `kustomization.yaml` files, which configure different components to implement our image updater GitOps workflow.

- `manifests`: Contains the necessary definitions to deploy your application.

- `namespace.yaml`: Creates a production namespace for your application.

- `helm-charts.yaml`: A definition to access your Helm chart.

- `helm-release.yaml`: Includes a definition and values to deploy your application using the Helm chart defined in `helm-charts.yaml`.

- `app-demo-registry.yaml`: Contains an image to scan on Docker Hub.

- `app-demo-policy.yaml`: Contains an expression to check inside files where you want to update the container image.

- `flux-system-automation.yaml`: Looks for a folder to update changes.

This repository is designed for your applications. It is a monorepo for a production environment with different clusters. You can do more complex configurations using Kustomize, but that is out of the scope of this chapter.

> **Important Note**
> You can also find some approaches to how to organize your repositories on the Flux website. For more information, check out the following link: `https://fluxcd.io/docs/guides/repository-structure`.

Now, it's time to see the workflow that we are going to implement in our GitOps use case for edge computing.

Understanding the application and GitHub Actions

To start implementing GitOps with Flux, we have to set a small pipeline that creates a container image every time we modify the source code of our application. To simplify our work, this configuration will be based on the `https://github.com/sergioarmgpl/fluxappdemo` repository, which contains a simple Python application using Flask. This application has two directories: `src` and `.github/workflows`. The `src` directory contains the source of the application, while the `workflows` folder has the GitHub Actions configuration.

So, let's first explore the `src` directory. You can see an overview of the repository in the following screenshot:

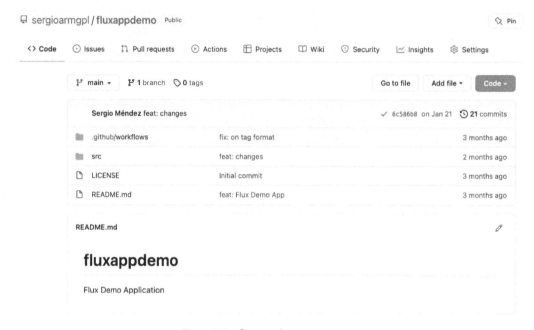

Figure 7.5 – fluxappdemo repository

The `src` directory contains the following files:

- `Dockerfile`: This has the configuration to build a Docker image; it also calls a small unit test included in `tests.py`.

- `Operations.py`: This has a class called `Operations` that contains a `runningInfo` method. This function receives two parameters: `msg1` and `msg2`. With these parameters, it returns the following message: `Running app <msg1> in namespace <msg2>`.

- `build_push.sh`: This is a sample script to build an image manually. It receives two parameters; the first one is your Docker username and the second is a tag for the image. You can run it as follows:

    ```
    $ /bin/bash build_push.sh <DOCKER_USERNAME> <IMAGE_TAG>
    ```

- `index.py`: This is the main Python file to run our application. It has a function called `hello_world` that gets the `MESSAGE` and `NAMESPACE` environment variables and then calls the `runningInfo` function to return the following message: `Running app <MESSAGE> in namespace <NAMESPACE>`. So, every time you call the application in `route /` and port `5000`, it will show the message, then `route /_health return Running` message, /_version a custom message. You could use this route to explore the application. To take a look at the code, check out the following link: `https://github.com/sergioarmgpl/fluxappdemo/blob/main/src/index.py`.

- `requirements.txt`: Includes all the necessary libraries to run the code.
- `tests.py`: This file includes a small test for the `runningInfo` function inside the `Operations` class.

You can see an overview of the `src` directory in the following screenshot:

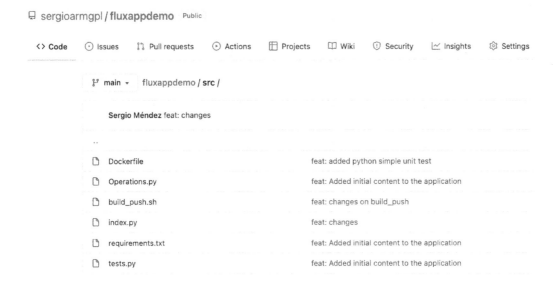

Figure 7.6 – src directory with source code

`.github/workflows` contains a `github-actions-fluxappdemo.yml` file. This file contains a CI pipeline definition that builds and pushes your container ARM image using the following name and tag format: `<DOCKER_USER>/fluxappdemo:RELEASE.YEAR-MONTH-DAYT-HOUR-MINUTE-SECONDZ`.

`DOCKER_USER` is your Docker username.

You can see this file in the following screenshot:

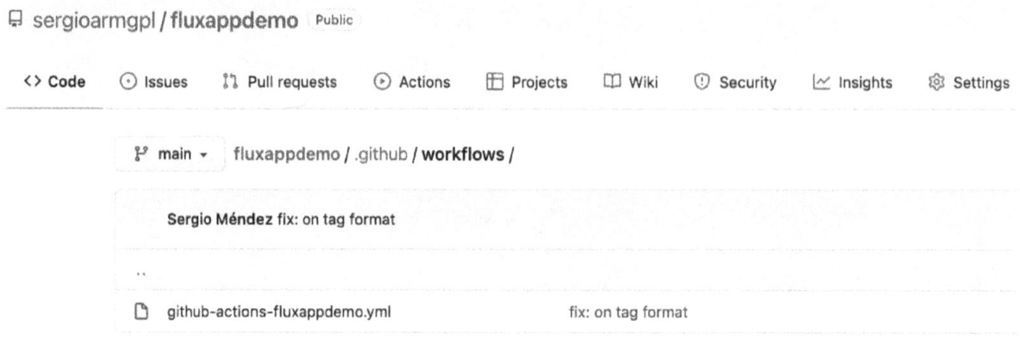

Figure 7.7 – GitHub Actions workflows file

With this brief explanation, let's move on to configure your own pipeline to build and push your container image.

Building your container image with GitHub Actions

To build and push your image with GitHub Actions, you should follow the given steps:

1. Fork the `https://github.com/sergioarmgpl/fluxappdemo` repository. This is going to create a repository named `https://github.com/<GITHUB_USER>/fluxappdemo`.

    `GITHUB_USER` is the username of your GitHub account. Replace it with your own username.

2. Create `DOCKERHUB_USERNAME` and `DOCKERHUB_TOKEN` secrets for your repository. These will be created as encrypted secrets for a repository. To create the secrets, open the following page in the browser: `https://github.com/<GITHUB_USER>/fluxappdemo/settings/secrets/actions`.

After adding the variables, your repository will look like this:

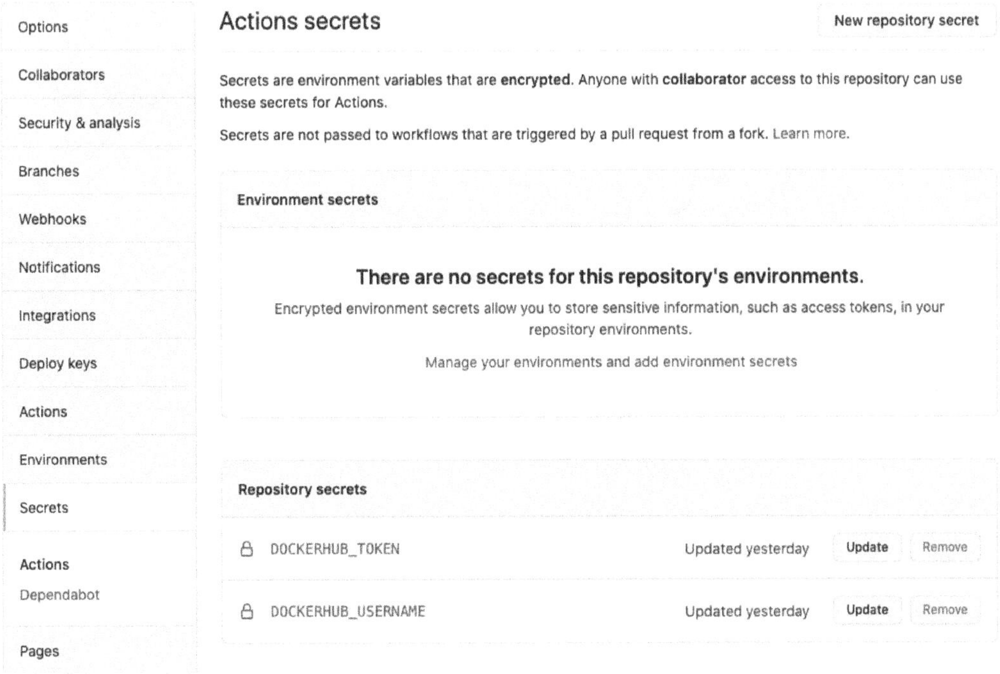

Figure 7.8 – GitHub repository secrets

3. Modify the `.github/workflows/github-actions-fluxappdemo.yml` file in the last line in the `tags` section with your user. It will look like this:

```
tags: <DOCKER_USER>/fluxappdemo
```

Here, DOCKER_USER is your Docker Hub username.

4. Commit and push the changes.

5. (*Optional*) To check whether your GitHub action is running, you can check out the following link: `https://github.com/<GITHUB_USER>/fluxappdemo/actions`. The following screenshot provides an example of how this should look:

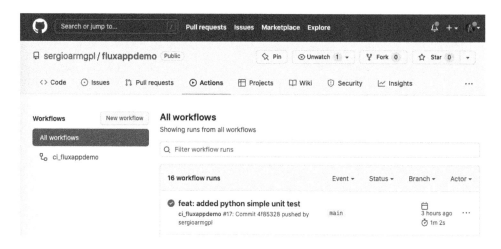

Figure 7.9 – GitHub Actions showing current workflows

6. (*Optional*) To check new container tags created for your account, check out the following link: `https://hub.docker.com/repository/docker/<DOCKERHUB_USERNAME>/fluxappdemo/tags`. The following screenshot provides an example of how this should look:

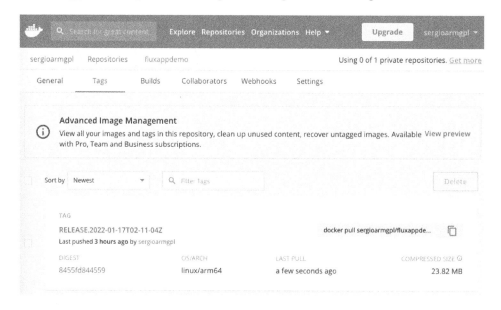

Figure 7.10 – Docker Hub tags for a repository

7. (*Optional*) To test whether your GitHub Actions pipeline works, modify the source code of the application inside the `src` directory and commit and push the changes. Then, a new workflow will be running.

> **Important Note**
>
> For more information about creating encrypted secrets for a repository, check out the following link: `https://docs.github.com/en/actions/security-guides/encrypted-secrets`. To create a token to access your Docker Hub account to push new images, check out this link: `https://docs.docker.com/docker-hub/access-tokens`. Finally, to fork a repository, check out the following link: `https://docs.github.com/en/get-started/quickstart/fork-a-repo`.

Now that we have configured a GitHub Actions pipeline to automate the creation of a container image with its tags, it's time to start configuring Flux to complete our GitOps workflow in the next section.

Installing and configuring Flux for GitOps

Before configuring Flux, let's understand what we are going to install in this section. In this section, we are going to install Flux and its components that detect new image tags for your container. Once new images are detected, Flux modifies the `HelmRelease` definition file inside your deployment repository. Then, Flux will automatically reconcile changes, updating the application deployment defined in this `HelmRelease` file that is using the Helm chart published at `https://sergiops.xyz/helm-charts`. Together with the GitHub Actions workflow defined in the *Building your container image with GitHub Actions* section, the complete workflow is going to work like this:

1. The user pushes changes from its local repository copy to the original source code repository located at `https://github.com/<GITHUB_USER>/fluxappdemo`.

2. GitHub Actions builds and pushes the image to Docker Hub at `https://hub.docker.com/repository/docker/<DOCKER_USER>/fluxappdemo`.

3. Flux detects the new tag generated when the image was updated.

4. Flux replaces the `HelmRelease` definition with the new tag. For this, Flux modifies, commits, and pushes the changes to the repository with your deployment definitions at `https://github.com/<GITHUB_USER>/fluxdemo-production.git`.

5. Flux reconciles the changes, and the application is updated with the new image tag.

> **Note**
>
> The `<GITHUB_USER>` and `<DOCKER_USER>` values have to be replaced with your GitHub and Docker users.

To start building this use case scenario with Flux, you have to install the Flux **CLI**. Here's how to do it:

1. To install the Flux CLI on Linux, run the following command:

    ```
    $ curl -s https://fluxcd.io/install.sh | sudo bash
    ```

2. Or, if you have macOS, you can install Flux with Homebrew using the following command:

    ```
    $ brew install fluxcd/tap/flux
    ```

> **Important Note**
> You can check for alternative installation at the official website, https://fluxcd.io.

Now, you have to install the Flux image updater feature, necessary to automate the CD process every time that a new image tag is detected. To install Flux and all necessary components, follow the given steps on each cluster:

1. Set your environment variables; in this case, we need to set our GitHub user and the token of our GitHub account, as follows:

    ```
    $ export GITHUB_USER=<YOUR_USER>
    $ export GITHUB_TOKEN=<YOUR_GITHUB_TOKEN>
    $ export DOCKER_USER=<YOUR_DOCKER_USERNAME>
    ```

> **Note**
> Check out the following link to create a token for your GitHub account: https://docs.github.com/en/authentication/keeping-your-account-and-data-secure/creating-a-personal-access-token.

2. Let's set the email address and username associated with your GitHub account. For this, run the following commands:

    ```
    $ git config --global user.email "<YOUR_EMAIL>"
    $ git config --global user.name "<YOUR_NAME>"
    ```

3. Now, install Flux and all the necessary components for image updater implementation. For this, run the following commands:

    ```
    $ CLUSTER_N=1

    $ flux bootstrap github \
    --kubeconfig /etc/rancher/k3s/k3s.yaml \
    ```

```
--components-extra=image-reflector-controller,image-
automation-controller \
--owner=$GITHUB_USER \
--repository=fluxdemo-production \
--branch=main \
--path=./clusters/cluster$CLUSTER_N \
--read-write-key \
--personal
```

The `repository` parameter is the name of the GitHub repository to create— for example, `fluxdemo-production`.

CLUSTER_N is an environment variable with the number of the cluster where you are installing Flux—for example, CLUSTER_1. The YAML files used to install Flux will be stored inside the `clusters/production/<CLUSTER_N>` directory.

The process will show the message **waiting for Kustomization "flux-system/flux-system" to be reconciled**. Once Flux is installed, you will see the message **all components are healthy**.

This command is going to create a repository with your user account. The link to access this repository will be `https://github.com/<GITHUB_USER>/fluxdemo-production.git`.

GITHUB_USER is the name of your GitHub username. Replace this value in the link with your own.

> **Important Note**
>
> Omit the `--components-extra=image-reflector-controller,image-automation-controller` line for `cluster2` and `cluster3`. `cluster1` is in charge of updating the application definitions for all clusters. Also refer to the help for this command by running `flux bootstrap github --help` for more options, especially if you are using an organization or enterprise or another versioning tool.

4. Clone the new repository and change it into this directory:

```
$ git clone https://github.com/$GITHUB_USER/fluxdemo-
production.git
$ cd fluxdemo-production
```

This is going to ask you to enter your username and password to clone your repository. This will be used in the next steps to customize and create deployment definitions. Proceed as follows:

1. Create a namespace for your application inside a directory called clusters/<clusterN>/ manifests. For this, run the following commands:

```
$ mkdir -p ./clusters/cluster$CLUSTER_N/manifests


$ kubectl create ns production --dry-run=client -o YAML >
./clusters/cluster$CLUSTER_N/manifests/namespace.yaml
```

Create a Helm chart source Flux object to point to your own Helm chart, as follows:

```
$ flux create source helm helm-charts \
--kubeconfig /etc/rancher/k3s/k3s.yaml \
--url=https://sergiops.xyz/helm-charts \
--interval=1m \
--namespace=production \
--export > ./clusters/cluster$CLUSTER_N/manifests/helm-
charts.yaml
```

In this example, we are using a Helm chart with a sample application in https:// sergiops.xyz/helm-charts.

2. Create a Flux HelmRelease object to create a YAML definition for your application deployment, as follows:

```
$ flux create helmrelease app-demo --chart app-demo \
--source HelmRepository/helm-charts.production \
--chart-version 0.0.1 \
--interval=1m \
--namespace production \
--export > ./clusters/cluster$CLUSTER_N/manifests/helm-
release.yaml
```

Let's add a section called values to the file by running the following command:

```
cat << EOF >> ./clusters/cluster$CLUSTER_N/manifests/
helm-release.yaml
  values:
    replicaCount: 3
    containerPort: 5000
    dockerImage: $DOCKER_USER/fluxappdemo:RELEASE.2022-
01-16T05-42-20Z # {"\$imagepolicy": "flux-system:app-
```

```
demo"}
    namespace: "production"
    domain: "app-demo-cluster$CLUSTER_N.domain.tld"
    changeCause: "First Deployment cluster $CLUSTER_N"
    message: "cluster$CLUSTER_N"
    appname: "app-demo-cluster$CLUSTER_N"
    node: "machine$CLUSTER_N"
EOF
```

Pay attention to the commented line `# {"$imagepolicy": "flux-system:app-demo"}`.

This part tells Flux to find where to replace the image with the new tag. Finally, the file will look like this:

```
apiVersion: helm.toolkit.fluxcd.io/v2beta1
kind: HelmRelease
metadata:
  name: app-demo
  namespace: production
spec:
  chart:
    spec:
      chart: app-demo
      sourceRef:
        kind: HelmRepository
        name: helm-charts
        namespace: production
      version: 0.0.1
  interval: 1m0s
  values:
    replicaCount: 3
    containerPort: 5000
    dockerImage: <DOCKER_USER>/fluxappdemo:RELEASE.2022-01-16T05-42-20Z # {"$imagepolicy": "flux-system:app-demo"}
    namespace: "production"
    domain: "app-demo-cluster1.domain.tld"
    changeCause: "First Deployment cluster 1"
```

```
        message: "cluster1"
        appname: "app-demo-cluster1"
        node: "machine1"
```

This `HelmRelease` object provided with the Flux installation provides a declarative way to parametrize a deployment. In this case, the values inside the `values` section correspond to the different parameters that you can send to our Helm chart. The creation of Helm charts is out of the scope of this book, but you can find good resources at the end of the chapter in the *Further reading* section.

Note

You can check the repository at `https://github.com/sergioarmgpl/fluxdemo-production` to see the final results of creating and modifying configuration files for your demo application with the previous commands.

This `HelmRelease` object is stored in a file and is the object that you need to modify, commit, and push your changes to your repository. After this, Flux detects the changes and updates your application. This file is inside of your repository in the `clusters/cluster$CLUSTER_N/manifests/helm-release.yaml` path. In this way, you can test how Flux updates your application, where the `CLUSTER_N` variable is the cluster number that you are modifying.

Important Note

If you want to create your own Helm chart repository, you can check out `https://helm.sh/docs/topics/chart_repository` and go to the *GitHub Pages example* section. In this example, we are using the charts located at `https://sergiops.xyz/helm-charts` and `https://github.com/sergioarmgpl/helm-charts/tree/gh-pages`. You can check or clone this repository to create your own.

You can omit Steps 11-13 if you are configuring cluster2 and cluster3 because cluster1 will be in charge of updating all the deployments' definitions.

3.   (*Optional*) Create an image repository to detect new releases or tags for your image, as follows:

```
$ flux create image repository app-demo \
--kubeconfig /etc/rancher/k3s/k3s.yaml \
--image=$DOCKER_USER/fluxappdemo \
--namespace=flux-system \
--interval=1m \
--export > ./clusters/cluster$CLUSTER_N/manifests/
app-demo-registry.yaml
```

4. (*Optional*) Create an image policy to define an expression to match to detect new image tags or releases from your image registry. This use case is going to use the following format to tag the new Docker images:

```
RELEASE.YEAR-MONTH-DAYT-HOUR-MINUTE-SECONDZ
```

This convention is based on RFC3339 and ISO 7601, which refers to the standards for date and time on the internet:

```
$ flux create image policy app-demo \
--image-ref=app-demo \
--namespace=flux-system \
--select-alpha=asc \
--filter-regex='^RELEASE\.(?P<timestamp>.*)Z$' \
--filter-extract='$timestamp' \
--export > ./clusters/cluster$CLUSTER_N/manifests/
app-demo-policy.yaml
```

5. (*Optional*) Now, it's time to put all the pieces together. For this, you have to create an `ImageUpdateAutomation` object, which is going to detect new releases and update the images in your deployments' YAML definitions. In the following case, it's going to check the folder clusters to update all the YAML definitions:

```
$ flux create image update flux-system \
--git-repo-ref=flux-system \
--git-repo-path="./clusters" \
--checkout-branch=main \
--push-branch=main \
--author-name=<AUTHOR_NAME> \
--author-email=<AUTHOR_EMAIL> \
--commit-template="{{range .Updated.Images}}{{println .}}
{{end}}" \
--export > ./clusters/cluster$CLUSTER_N/manifests/flux-
system-automation.yaml
```

You have to change the <AUTHOR_NAME> and <AUTHOR_EMAIL> tags with your own values. This will appear as the commit author when Flux pushes changes for image tags.

6. Commit and push the changes to the repository with the following commands:

```
$ git add -A
$ git commit -m "feat: App YAML definitions"
$ git push origin main
```

The `push` command is going to ask you for the user and the token that you previously created, to access your GitHub account.

Now, you can build a new image, and you can wait for Flux to automatically update your `HelmRelease` file with the new image detected. After 1 minute or more, you can expect the change to have been made. You will expect to see a commit in your repository made by Flux with the new tag detected, to troubleshoot whether the image updater is working.

7. (*Optional*) You can force Flux to apply this configuration by running the Flux reconciliation process with the following command:

    ```
    $ flux reconcile kustomization flux-system --with-source
    --kubeconfig /etc/rancher/k3s/k3s.yaml
    ```

8. The Helm chart is going to provision a `LoadBalancer` service type. To find the provisioned IP address, run the following command:

    ```
    $ IP_LOADBALANCER=$(kubectl get svc app-demo-
    cluster$CLUSTER_N-srv --output jsonpath='{.status.
    loadBalancer.ingress[0].ip}' -n production)
    ```

    Here, the `IP_LOADBALANCER` variable contains the IP of the load balancer created by the `HelmRelease` definition, which is the endpoint for your application in this cluster. You can check the value by running the following command:

    ```
    $ echo $IP_LOADBALANCER
    ```

    Using *Figure 7.4* as a reference, you will expect to see an IP address such as `192.168.0.52`.

9. Let's say that, for example, the returned IP is `192.168.0.52`. You can access your application with the following URL: `http://192.168.0.52:5000`. You can test to access other routes—for example, `/_version` or `/_health`.

Now that you have installed Flux, you can start testing the auto-reconciliation to update your files by committing and pushing the changes of your `HelmRelease` files. The auto-reconciliation updates everything Flux detects in new image tags of your applications. This process is described in the previous section, *Designing GitOps with Flux for edge applications*. After this, you can continue with the next section to learn how to troubleshoot your installation.

Troubleshooting Flux installations

There are a few useful commands that can help you to troubleshoot your installation; in this section, we're going to find out what these are. So, let's proceed as follows:

1. To reconcile Flux changes in Flux, run the following command:

    ```
    $ watch flux get images all --all-namespaces --kubeconfig
    /etc/rancher/k3s/k3s.yaml
    ```

    This command is going to show new tags detected for your container, and how these new tags are set up in your HelmRelease YAML definition file.

2. To check the image repositories in Flux, run the following command:

    ```
    $ flux get image repository app-demo --kubeconfig /etc/
    rancher/k3s/k3s.yaml --namespace=production
    ```

3. To check the current policy in your cluster, run the following command:

    ```
    $ flux get image policy app-demo --kubeconfig /etc/
    rancher/k3s/k3s.yaml --namespace=production
    ```

4. To get all images configured in your Flux installation, run the following command:

    ```
    $ flux get images all --all-namespaces --kubeconfig /etc/
    rancher/k3s/k3s.yaml
    ```

5. To reconcile YAML definition changes in your cluster, run the following command:

    ```
    $ flux reconcile kustomization flux-system --with-source
    --kubeconfig /etc/rancher/k3s/k3s.yaml
    ```

6. To watch in real time how image detection and updates to your repositories are running, run the following command:

    ```
    $ watch flux get images all --all-namespaces --kubeconfig
    /etc/rancher/k3s/k3s.yaml
    ```

7. To check your application deployments, run the following command:

    ```
    $ kubectl get deploy -n production
    ```

8. To check your Pods, run the following command:

    ```
    $ kubectl get pods -n production
    ```

You have now learned these essential commands to troubleshoot your Flux system.

In the next section, we are going to explore Flux monitoring dashboards.

Installing Flux monitoring dashboards

Flux itself doesn't include a graphical user interface for management but integrates some useful dashboards using Prometheus and Grafana to visualize the state of your deployments. These dashboards have to be installed on each cluster. To install this feature, follow the next steps:

1. Configure the Git repository that contains monitoring stack definitions for its installation. The configuration will listen for changes every 30 minutes. The code is illustrated here:

    ```
    $ flux create source git monitoring \
    --interval=30m \
    --kubeconfig /etc/rancher/k3s/k3s.yaml \
    --url=https://github.com/fluxcd/flux2 \
    --branch=main
    ```

2. Install kube-prometheus-stack, which is going to be used to configure Prometheus for your dashboards. This stack will be installed in the monitoring namespace. The code is illustrated in the following snippet:

    ```
    $ flux create kustomization monitoring-stack \
    --interval=1h \
    --kubeconfig /etc/rancher/k3s/k3s.yaml \
    --prune=true \
    --source=monitoring \
    --path="./manifests/monitoring/kube-prometheus-stack" \
    --health-check="Deployment/kube-prometheus-stack-
    operator.monitoring" \
    --health-check="Deployment/kube-prometheus-stack-grafana.
    monitoring" \
    --health-check-timeout="5m0s"
    ```

3. Install Grafana and configure your Flux dashboards, storing data in Prometheus and visualizing this across preconfigured dashboards in Grafana. The code is illustrated in the following snippet:

    ```
    $ flux create kustomization monitoring-config \
    --interval=1h \
    --kubeconfig /etc/rancher/k3s/k3s.yaml \
    --prune=true \
    ```

```
--source=monitoring \
--path="./manifests/monitoring/monitoring-config"
```

4. Access the dashboards using the next command:

    ```
    $ kubectl -n monitoring port-forward svc/kube-prometheus-
    stack-grafana --address 0.0.0.0 3000:80
    ```

    This is going to open port 3000 of your dashboard. Remember that the IP address that you have to access is the IP of the node where you are accessing this dashboard.

    Access the dashboard using the following URL: http://<NODE_IP_ADDRESS>:3000/d/flux-control-plane.

    NODE_IP_ADDRESS is the IP address cluster node where you are running the command shown in this step.

5. To access the dashboard, use the next credentials:

    * Username: admin

    * Password: prom-operator

    The login screen will look like this:

Figure 7.11 – Grafana login form

Once you are logged in, you will be redirected to the dashboard URL previously mentioned.

6.   Once the dashboard is opened, it will look like this:

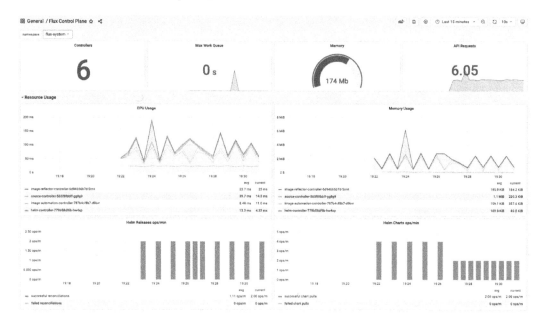

Figure 7.12 – Grafana Flux control plane dashboard

> **Important Note**
>
> Remember that you can customize this dashboard and create your own dashboards. For more information about this dashboard, you can visit the following link: `https://fluxcd.io/docs/guides/monitoring`.

Your Grafana dashboards are now installed successfully, and you can see the state of your deployments. Now, it's time to learn how to uninstall Flux in the next section.

Uninstalling Flux

Once you don't need the Flux installation anymore, you can run the following command:

```
$ flux uninstall -s --namespace=flux-system --kubeconfig /etc/
rancher/k3s/k3s.yaml
```

This is going to uninstall Flux from your Kubernetes cluster. Now, it's time to finish the chapter.

Summary

In this chapter, we learned how GitOps works and how you can implement GitOps using GitHub Actions and Flux. Flux could be useful to automate your deployments on an edge environment, using a single Git repository. For this, we learned how Flux can implement GitOps to update your applications at the edge using the HelmRelease object and the image updater feature. Flux can manage your application cluster without using an external way to expose the Kubernetes API of your cluster, which is the case with tools such as Argo CD. This can be translated into cost savings and a more effective tool for edge environments using ARM devices. Argo CD, on the other hand, doesn't support ARM and needs a way to expose your Kubernetes API from your cluster to connect the cluster to Argo CD using a public load balancer or a virtual machine on the internet. In the next chapter, we are going to learn how to add basic observability and traffic splitting to your applications using Linkerd.

Questions

Here are a few questions to validate your new knowledge:

- How can GitHub Actions help me to implement GitOps at the edge?
- How can I implement GitOps using Flux?
- Which other features does Flux have that can help me to implement GitOps?
- How can I troubleshoot my Helm releases with Flux?
- How can I apply this simple use case in my edge computing scenarios?
- How can I structure my repositories for GitOps?

Further reading

You can refer to the following references for more information on the topics covered in this chapter:

- *What is GitOps?*: https://www.gitops.tech
- *Dev Leaders Compare Continuous Delivery vs Continuous Deployment vs Continuous Integration*: https://stackify.com/continuous-delivery-vs-continuous-deployment-vs-continuous-integration
- Quickstart on using GitHub: https://docs.github.com/en/get-started/quickstart
- GitHub Actions to build container images: https://github.com/docker/build-push-action
- Docker Hub: https://hub.docker.com

- Creating secrets on GitHub Actions: `https://docs.github.com/en/actions/security-guides/encrypted-secrets`

- Date and time internet standard: `https://datatracker.ietf.org/doc/html/rfc3339`

- Creating a public Helm chart with GitHub Pages: `https://medium.com/@mattiaperi/create-a-public-helm-chart-repository-with-github-pages-49b180dbb417`

- Creating a Helm chart repository: `https://harness.io/blog/helm-chart-repo`

- How to structure your Flux repositories: `https://fluxcd.io/docs/guides/repository-structure`

- *Flux Documentation*: `https://fluxcd.io/docs`

- Flux Helm releases: `https://fluxcd.io/docs/guides/helmreleases`

- Flux, Kustomize, and Helm example: `https://github.com/fluxcd/flux2-kustomize-helm-example`

8
Observability and Traffic Splitting Using Linkerd

Observability is important when you develop microservices or applications using containers, as it provides insights into complex systems. Monitoring mechanisms, analytics, and observability give you an idea of how your applications will work in production as a system. In production, observability provides logging, metrics, and traces of how services interact with one another to provide functionality. Service meshes are often used to implement observability in your services. A **service mesh** is a powerful tool that helps you to implement observability and other functionalities such as retries or timeout management, without modifying your applications. This chapter discusses **golden metrics**, commonly used metrics for understanding systems, how to implement observability using Linkerd for an application with an ingress controller, and how to implement traffic routing using a sample application.

In this chapter, we're going to cover the following main topics:

- Observability, monitoring, and analytics
- Introduction to service meshes and Linkerd
- Implementing observability and traffic splitting with Linkerd
- Testing observability and traffic splitting with Linkerd
- Uninstalling Linkerd
- Ideas to implement using service meshes

Technical requirements

In this chapter, to implement observability with Linkerd, you will need the following:

- A single or multi-node K3s cluster using ARM devices with MetalLB installed and with the option to avoid Traefik being installed as the default ingress controller.

- Kubectl configured to be used in your local machine to avoid using the `--kubeconfig` parameter.

- Helm command installed.

- Clone the repository at `https://github.com/PacktPublishing/Edge-Computing-Systems-with-Kubernetes/tree/main/ch8` if you want to run the YAML configuration by using `kubectl apply` instead of copying the code from the book. Take a look at the `yaml` directory for the YAML examples, inside the `ch8` directory.

We are going to install Linkerd to implement observability and traffic splitting on this cluster. So, let's get started with the basic theory to understand the benefits of observability and how to implement it.

Observability, monitoring, and analytics

To start, let's get familiar with the observability concept. Peter Waterhouse mentioned, in his article in *The New Stack*, that "*observability is a measure of how well internal states of a system can be inferred from knowledge of its external outputs.*" He also mentioned that observability is more of a property of a system and not something that you actually do.

There are two concepts that are close to each other in this context: monitoring and observability. In Steve Waterworth's article, available at `dzone.com`, he mentioned this relation with the phrase, "*If you are observable, I can monitor you.*"

What this means is that observability is achieved when data about systems is managed. Monitoring, on the other hand, it is the actual task of collecting and displaying this data. Finally, the analysis occurs after collecting data with a monitoring tool, and you perform it either manually or automatically.

This relationship is represented by the Pyramid of Power:

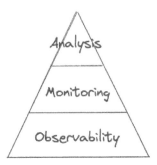

Figure 8.1 – Pyramid of Power

The Pyramid of Power represents how analysis and monitoring are the base to implement observability. Together, they can bring the property to know the state of your system; this is what we call observability. Service meshes give observability to the system by measuring metrics that reflect the state of the system. These metrics are called golden metrics. Let's explore golden metrics in the next section.

Golden metrics

Golden metrics were first introduced in the Google *Site Reliability Engineering* book and were defined as the minimum metrics required to monitor services. This is how the Pyramid of Power gets a place in the discussion about monitoring and observability. These metrics were also defined as a model, as a foundation for building monitoring around applications.

According to the Linkerd service mesh glossary web page, golden metrics are also called **golden signals**; these are the core metrics of application health. These metrics are defined or based on latency, traffic volume, error rate, and saturation. With these metrics, you can figure out the health of your application to finally build the property of observability in your applications and system. Golden metrics are the base for monitoring services and building observable systems.

Let's explore, in the next section, how service meshes implement these golden metrics to bring observability to your system.

Introduction to service meshes and Linkerd

George Mirando, in his book, *The Service Mesh*, says that a service mesh "*is a dedicated infrastructure layer for handling service-to-service communication in order to make it visible, manageable, and controlled. The exact details of its architecture vary between implementations, but generally speaking, every service mesh is implemented as a series of interconnected network proxies designed to better manage service traffic.*" In general, we can adopt the idea of a service mesh being built by this interconnected network of proxies that provides manageable, stable, and controlled service-to-service communication.

Now, let's see how this is implemented, starting with the explanation given in the following diagram:

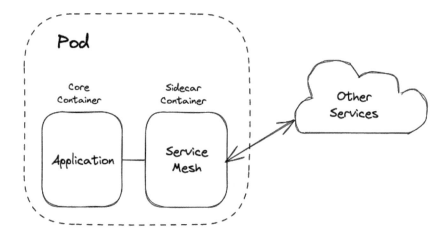

Figure 8.2 – Service mesh implementation with a sidecar container

Sidecar is a design pattern used on distributed systems that only have a single node. This pattern is commonly used in Kubernetes when deploying applications that use multiple containers. In this context, the sidecar pattern is made with two containers; the first container contains the application container (which is the core container), and the second sidecar container is a proxy that provides functionalities for a reliable network for your application, and both live inside a pod (which is an abstraction of a group of containers in Kubernetes for an application). This pod lives inside a data plane that contains all the services interconnected by proxies. To exemplify this, let's look at the following diagram:

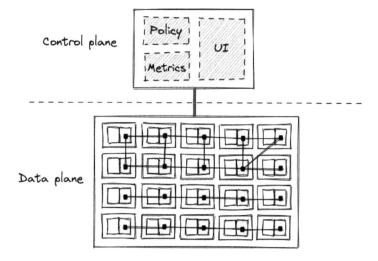

Figure 8.3 – Service mesh control plane and data plane

These proxies ask the control plane what to do with the incoming traffic, for example, block or encrypt the traffic. The control plane also evaluates and decides the corrective action to run in the proxies, such as a retry or redirect if a timeout occurs. The control plane contains rules to be applied to each service connect across the mesh. Collecting data to provide golden metrics makes the services observable. Some service meshes also provide a basic UI to manage all these service mesh functionalities.

The need for service meshes exists because of wrong assumptions regarding distributed systems, such as the following:

- The network is reliable.

- Latency is zero.

- Bandwidth is infinite.

- The network is secure.

- Topology doesn't change.

- There is one administrator.

- The transport cost is zero.

- The network is homogeneous.

Service meshes exist to address all the wrong assumptions, helping to manage distributed systems from the logic in your application code and creating a reliable network for your application. In general, service meshes provide this reliability by just injecting a proxy as a sidecar without modifying the code of your application.

Finally, the relationship between service meshes and observability is that these proxies can generate the golden metrics when the proxies intercept network traffic, providing a graphical dashboard to provide a way to visualize the state of your applications; in other words, creating the observability property for your system.

Linkerd service mesh

Linkerd is a service designed to run on Kubernetes. It provides debugging, observability, reliability, and security to your applications deployed on Kubernetes without modifying your application's source code. So, Linkerd not only provides observability but also provides more features, such as the following:

- HTTP, HTTP/2, and gRPC proxying

- Retries and timeouts

- Telemetry and monitoring

- Load balancing

- Authorization policy

- Automatic proxy injection

- Distributed tracing

- Fault injection

- Traffic split

- Service profiles

- Multi-cluster communication

Linkerd is also a fully open source software, part of the graduated projects of the **Cloud Native Computing Foundation (CNCF)**. Linkerd is built by Buoyant.

As we explored in the introduction to service meshes, Linkerd works with a data plane and a control plane, and it has the Linkerd CLI to manage its installation. It also comes with a UI to explore the different graphics that show golden metrics for your injected services.

In order to use Linkerd, first, you have to inject your application with the Linkerd proxy using the Linkerd CLI, and then Linkerd will be ready to start collecting metrics and enable your application to communicate with other inject services across the data plane; and, of course, Linkerd will be ready to configure your application with all its features such as traffic splitting.

Linkerd was designed to be fast without consuming a lot of resources and to be easy to use compared to other service meshes such as **Istio**. Istio includes a full package of tools for implementing not only a service mesh functionality but also tracing and ingress controller functionalities, which could be too much for some solutions. Linkerd reduces the complexity, and it was built to work as a modular service mesh piece of software that can integrate with your current technology solution stack to add an observability layer to your system. Linkerd meets edge computing requirements supporting ARM architectures and low resource consumption and is simple to use. In this way, Linkerd could be an option to look at before considering another solution based on Envoy such as Istio.

It's important to mention that, because service meshes work using proxies, some ingress controllers or cloud native proxies could match your needs before choosing a full service mesh solution such as Traefik, Emissary, and Contour. Some important features to consider while picking a service mesh or a cloud native proxy are security and rate limit implementations. You can explore some articles comparing these solutions in the *Further reading* section. But now, it's time to understand how to implement observability and traffic splitting in the next section.

Implementing observability and traffic splitting with Linkerd

To explain how we are going to use Linkerd for observability and traffic splitting, let's explore the following diagram:

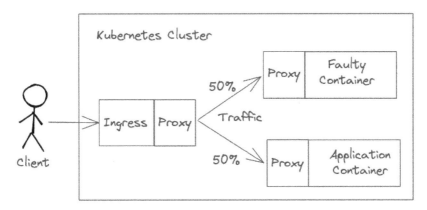

Figure 8.4 – Traffic splitting with Linkerd

First of all, you have to install Linkerd in your Kubernetes cluster. For this small scenario, we are going to use two deployments. The first deployment is a simple API deployment that returns the message *Meshed application app1 with Linkerd*, and the second, a deployment that always returns error code 500.

All the traffic will be sent by a client (in our case a loop that sends requests to the endpoint of the application) that is a load balancer created by your ingress controller service and used by an ingress definition. Every time the ingress object detects the traffic, the traffic will be split by 50% to the API deployment and 50% to the faulty deployment. This is going to simulate an error rate of 50% in your requests and 50% for traffic without errors.

It's necessary to inject the ingress, the application, and the faulty deployment that simulates errors. In this way, these services will communicate with each other using the Linkerd proxy injected on each deployment.

While the traffic is moving across the services, it is generating the golden metrics that the Linkerd dashboard can visualize with Grafana and other reports that Linkerd implements in its UI.

Now, we are ready to start installing Linkerd in the next section.

Installing Linkerd in your cluster

So, let's begin with the installation of Linkerd in your cluster. For this you have to follow the next steps:

1. First, install the Linkerd CLI by running the following command:

    ```
    $ curl --proto '=https' --tlsv1.2 -sSfL https://run.
    linkerd.io/install | sh
    ```

    If you are using macOS, you can install the Linkerd CLI using the brew command:

    ```
    $ brew install linkerd
    ```

2. Add the directory where Linkerd is installed to your path:

    ```
    $ echo  "export PATH=\$PATH:/home/ubuntu/.linkerd2/bin"
    >> ~/.bashrc
    ```

    Run the following command to load the new path, instead of logging in again to load the new path:

    ```
    $ source ~/.bashrc
    ```

3. To check whether the cluster fits the requirements to install Linkerd, run the following:

    ```
    $ linkerd check --pre
    ```

4. Next, install Linkerd by running the following command:

    ```
    $ linkerd install | kubectl apply -f -
    ```

5. Now, install the Linkerd dashboard by running the following command:

    ```
    $ linkerd viz install | kubectl apply -f -
    ```

    This command is going to wait while Linkerd is being installed before installing the Linkerd dashboard.

6. To check whether the installation was successful, run the following:

    ```
    $ linkerd check
    ```

7. To open the Linkerd dashboard once everything is running, run the following command:

    ```
    $ linkerd viz dashboard --address 0.0.0.0
    ```

The previous command will expose the Linkerd dashboard inside your device. To run this command, we are assuming that the command was run inside the devices, so you need to run the following line to resolve the URL `http://web.linkerd-viz.svc.cluster.local:50750` to point to your device:

```
$ IP_CLUSTER=<YOUR_IP_CLUSTER>
$ sudo echo $IP_CLUSTER" WEB.linkerd-viz.svc.cluster.local" >>
/etc/hosts
```

`IP_CLUSTER` is the IP address of your cluster.

Now, access the next URL to open the dashboard: `http://web.linkerd-viz.svc.cluster.local:50750`.

Now, it's time to install the NGINX ingress controller to be used in this implementation. Let's explore this in the next section.

Installing and injecting the NGINX ingress controller

In this scenario, we are going to use the NGINX ingress controller, using Helm to install it by following the given steps:

1. Create the `nginx-ingress` namespace:

   ```
   $ kubectl create ns nginx-ingress
   ```

2. Add the NGINX ingress controller Helm chart and update the repositories configured in Helm:

   ```
   $ helm repo add ingress-nginx https://kubernetes.github.
   io/ingress-nginx
   $ helm repo update
   ```

3. Install the NGINX ingress controller:

   ```
   $ helm install nginx-ingress ingress-nginx/ingress-nginx
   -n nginx-ingress
   ```

4. Now, to inject the NGINX ingress controller pod, run the following command:

   ```
   $ kubectl get -n nginx-ingress deploy nginx-ingress-
   ingress-nginx-controller -o yaml \
   | linkerd inject - \
   | kubectl apply -f -
   ```

Your ingress controller is now ready to be installed and injected. Let's create the applications that we need in the next section.

Creating a demo application and faulty pods

Now, let's create our sample application and faulty pod to experiment with the traffic splitting feature and get some faulty traffic to simulate error requests. For this, follow the given steps:

1. Create the `myapps` namespace for your pods:

    ```
    $ kubectl create ns myapps
    ```

2. Create the sample application, `app1`, by running the following command:

    ```
    $ cat <<EOF | linkerd inject - | kubectl apply -f -
    apiVersion: apps/v1
    kind: Deployment
    metadata:
      labels:
        app: app1
      name: app1
      namespace: myapps
    spec:
      replicas: 1
      selector:
        matchLabels:
          app: app1
      template:
        metadata:
          labels:
            app: app1
        spec:
          containers:
          - image: czdev/app1demo
            name: app1demo
            env:
            - name: MESSAGE
              value:  "Meshed application app1 with Linkerd"
            - name: PORT
              value:  "5000"
    EOF
    ```

> **Important Note**
>
> The `linkerd inject` command inserts the `linkerd.io/inject: enabled` label in the `annotations` sections of your deployment or pod. This label is used by Linkerd to inject the services with the Linkerd proxy. You can also add this label manually in your YAML definitions to have a better approach using declarative definitions for your pods and deployments. To customize the code of app1demo check the link `https://github.com/sergioarmgp1/containers/tree/main/app1demo`.

3. To create our faulty pod, we are going to use NGINX as a web server and a custom configuration to return a request with a 500 code error in order for Linkerd to detect and count the request as an error. For this, let's create the configuration by running the following command:

```
$ cat <<EOF | kubectl apply -f -
apiVersion: v1
kind: ConfigMap
metadata:
  name: error-injector
  namespace: myapps
data:
nginx.conf: |-
    events {}
    http {
        server {
          listen 5000;
            location / {
                return 500;
            }
        }
    }
EOF
```

4. Now, let's create the deployment that returns a 500 error in port 5000 when accessing the pod in the / path:

```
$ cat <<EOF | linkerd inject - | kubectl apply -f -
apiVersion: apps/v1
kind: Deployment
metadata:
  name: error-injector
```

```
        namespace: myapps
        labels:
          app: error-injector
    spec:
      selector:
        matchLabels:
          app: error-injector
      replicas: 1
      template:
        metadata:
          labels:
            app: error-injector
        spec:
          containers:
            - name: nginx
              image: nginx:alpine
              volumeMounts:
                - name: nginx-config
                  mountPath: /etc/nginx/nginx.conf
                  subPath: nginx.conf
          volumes:
            - name: nginx-config
              configMap:
                name: error-injector
EOF
```

5. Now that our applications have been deployed, let's configure the services for these applications. Let's start with the error-injector service:

```
$ cat <<EOF | kubectl apply -f -
apiVersion: v1
kind: Service
metadata:
  name: error-injector
  namespace: myapps
spec:
  ports:
```

```
      - name: service
        port: 5000
      selector:
        app: error-injector
EOF
```

6. Now, create the service for your application by running the following command:

```
$ cat <<EOF | kubectl apply -f -
apiVersion: v1
kind: Service
metadata:
  name: app1
  namespace: myapps
spec:
  ports:
  - name: service
    port: 5000
  selector:
      app: app1
EOF
```

7. Now, let's use the **Service Mesh Interface (SMI)** specification to configure the traffic splitting. With this configuration, the traffic will be split by 50% to the app1 service and the other half for error-injector, so we are going to expect a 50% success rate:

```
$ cat <<EOF | kubectl apply -f -
apiVersion: split.smi-spec.io/v1alpha1
kind: TrafficSplit
metadata:
  name: error-split
  namespace: myapps
spec:
  service: app1
  backends:
  - service: app1
    weight: 500m
  - service: error-injector
```

```
        weight: 500m
    EOF
```

8. Finally, let's create our ingress rule to expose the endpoint to send traffic to this application using traffic splitting:

```
$ cat <<EOF | kubectl apply -f -
apiVersion: networking.k8s.io/v1
kind: Ingress
metadata:
  name: ingress
  namespace: myapps
  annotations:
    nginx.ingress.kubernetes.io/rewrite-target: /
    nginx.ingress.kubernetes.io/service-upstream:   "true"
spec:
  ingressClassName: nginx
  rules:
  - http:
      paths:
      - path: /
        pathType: Prefix
        backend:
          service:
            name: app1
            port:
              number: 5000
    EOF
```

> **Important Note**
>
> Depending on which Kubernetes version you are using, you have to use the syntax for your ingress controller definition, *v1beta1* or *v1*. For more information, you can check `https://kubernetes.io/docs/concepts/services-networking/ingress`, and change from different Kubernetes versions.

Now, we are ready to test the observability and traffic splitting configured with Linkerd. Let's explore this in the next section.

Testing observability and traffic splitting with Linkerd

Now, it's time to test the observability. To start exploring the dashboard and see the observability, follow the given steps:

1. Open your dashboard by running the following command:

```
$ linkerd viz dashboard
```

This will automatically open the dashboard to the URL `http://localhost:50750`.

The dashboard will look as in the following screenshot:

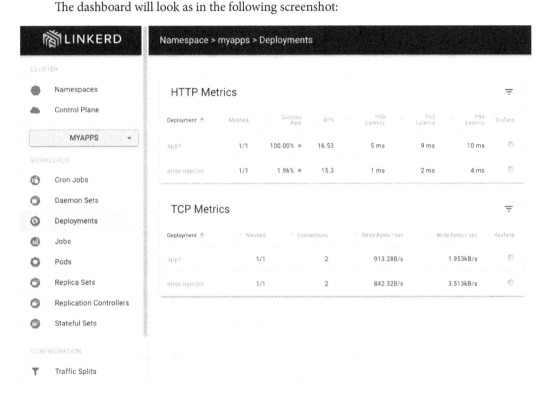

Figure 8.5 – Linkerd dashboard

To load the right information, select the **MYAPPS** namespace in the combo box in the left sidebar, and then click on the **Deployments** icon to load the **HTTP Metrics** and **TCP Metrics** information.

To see similar information as the previous dashboard, execute the following command to start sending traffic to our deployment:

```
$ ENDPOINT=$(kubectl get svc nginx-ingress-ingress-nginx-
controller --output jsonpath='{.status.loadBalancer.
```

```
ingress[0].ip}' -n nginx-ingress)
$ while true; do curl http://$ENDPOINT;echo  " "; done
```

The first command assigns the load balancer IP address that your NGINX ingress controller is using as the endpoint to expose services using ingress definitions. Then, while the command sends traffic, it also shows the result of each request, showing a similar message to the following:

```
Host:app1-555485df49-rjf4vMeshed application app1 with
Linkerd
```

Or, the following output error is displayed:

```
<html>
<head><title>500 Internal Server Error</title></head>
<body>
<center><h1>500 Internal Server Error</h1></center>
<hr><center>nginx/1.21.6</center>
</body>
</html>
```

This is a frequency of 50% for the message and 50% for the error, on average.

2. If you click on the orange Grafana icon (let's say, for example, in the **HTTP Metrics** section), you will see a similar Grafana graph to the following:

Figure 8.6 – Grafana Linkerd HTTP Metrics graph

In this graph, you can see the golden metrics and the success rate of the application for the `app1` deployment, the **requests per second** (**RPS**), and the latency of each request; these metrics represent the golden metrics for your application, which give you the basic observability feature for your system and your application.

3. If you click on **Traffic Splits** while the **myapps** namespace is selected, you will see a traffic splitting representation like this:

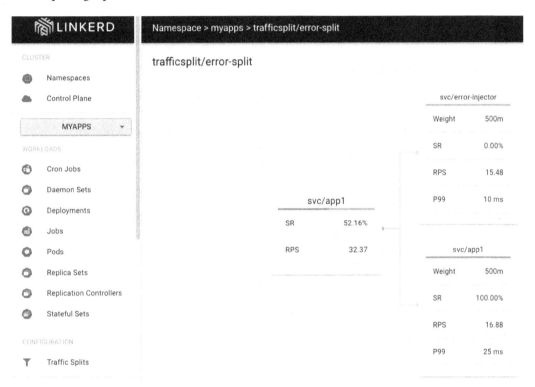

Figure 8.7 – Linkerd traffic splitting dashboard

In this dashboard, you will see, in real time, how the traffic splitting configuration sends 50% of the traffic to the `app1` Kubernetes services and 50% to the error injector. The red color represents failure requests (requests that return the `500` request error code), while the green color represents valid traffic from the `app1` service returning a `200` request code. This, in general, gives you the live state of your application, which is the goal of implementing observability.

This basic implementation simulates a failure request for an application using a service mesh. You can also use the same implementation to split your traffic between applications or implement advanced deployment strategies such as blue/green deployments. This was a simple use case to implement observability in your applications and the power of traffic management from a service mesh. Now, let's explore some useful commands if you want to use Linkerd using the CLI.

Using Linkerd's CLI

In some cases where no UI is available, maybe for security reasons, it could be useful to use the Linkerd CLI. So, let's explore four basic command-line options: `routes`, `top`, `tap`, and `edges`:

- `routes` shows the current routes that other applications or clients are using to access your application. Using our previous scenario as an example, you can show the routes of `app1` in the `myapps` namespace with the following command:

  ```
  $ linkerd viz routes deployment/app1 --namespace myapps
  ```

- `top` displays the traffic and path of your application. The following command is going to show how the ingress controller forwards the traffic to your applications, shows a counter to access the / path, and shows the success rate of the requests:

  ```
  $ linkerd viz top deployment/app1 --namespace myapps
  ```

- `tap` displays the information of the requests in real time for `app1`; for this, you have to run the following command:

  ```
  $ linkerd viz tap deployment/app1 --namespace myapps
  ```

- `edges` shows a table displaying how your application is connected with other injected applications in your cluster, and the source and destiny of each connection. For this, you have to run the following command for `app1`:

  ```
  $ linkerd viz edges po -n myapps
  ```

With this, you have an idea of how to use Linkerd with the CLI. Now, let's move to the next section to learn how to uninstall Linkerd.

Uninstalling Linkerd

If you are evaluating Linkerd or doing some management in your clusters, for example, it could be useful to uninstall Linkerd. For this, follow the next steps:

1. Uninstall support for additional features of Linkerd (called **viz**) as follows:

   ```
   $ linkerd viz uninstall | kubectl delete -f -
   ```

2. Uninstall the Linkerd control plane. This is going to uninstall the rest of the core Linkerd components. For this, run the following command:

   ```
   $ linkerd uninstall | kubectl delete -f -
   ```

Now, Linkerd is uninstalled from your cluster. To end this chapter, let's move to the last section to explore some useful ideas of where you can use Linkerd.

Ideas to implement when using service meshes

To end this chapter, here are some ideas of how you can get the advantages of using service meshes at the edge. These ideas are not specific to the edge and could be used in a common infrastructure:

- **Implement rate limits**: You can use a service mesh to configure some rate limits in your applications, managing in this way how much input traffic is accepted. There are some awesome projects to implement this, including Linkerd and Envoy-based service meshes such as Istio and Ambassador.

- **Traffic splitting**: You can use this feature of service meshes to implement blue/green deployments and canary deployments; an example of this is the implementation of Argo Rollouts, which can use Linkerd to implement this kind of deployment strategy. You can also implement some chaos engineering tests using service meshes.

- **Security policies**: You can use service meshes to restrict traffic and encrypt end-to-end traffic. This could be useful to increase the security of your services.

- **Multi-cluster connection**: With a service mesh, you can connect your clusters without complex configurations. **Kuma** is a control plane for microservices and service meshes that can help you to connect multiple clusters; it was built on top of Envoy. You can also do the same using Linkerd and other Envoy-based service meshes.

- **Scaling based on networking**: You can use Prometheus metrics generated by service meshes to generate alerts or scale your services. You can also implement machine learning models to implement some intelligent scaling. You can use them with projects such as **Kubernetes-based Event-Driven Autoscaling** (**KEDA**), which reads information from an API to scale your services.

These are some ideas that you can explore when using service meshes. Now, it's time to finish the chapter.

Summary

In this chapter, we learned how to implement observability and how to use a service mesh to set up traffic splitting. We focused on implementing this scenario using Linkerd, running a sample application that shows a message, and using traffic splitting. When the application receives the traffic, we showed how to explore the different graphics that can be used to get the real-time state of your system. We also learned how to use Linkerd with the CLI uninstalled. The chapter ended with some implementation ideas to explore when using service meshes and how this can impact your system. All of this forms the base to implement observability and basic traffic splitting in systems using a Linkerd service mesh. In the next chapter, we are going to learn how to implement serverless functions and simple event-driven pipelines using Knative.

Questions

Here are a few questions to validate your new knowledge:

- How do service meshes help you to implement observability?

- What are the features that service meshes provide to systems?

- How do service meshes work internally?

- What does Linkerd provide for users implementing observability?

- How can Linkerd be compared to other service meshes?

- What are the common use cases for service meshes?

Further reading

You can refer to the following resources for more information on the topics covered in this chapter:

- *Design distributed systems* book, by *Brendan Burns*: `https://learning.oreilly.com/library/view/designing-distributed-systems/9781491983638`

- Service mesh pattern: `https://philcalcado.com/2017/08/03/pattern_service_mesh.html`

- *Golden Signals - Monitoring from first principles*: `https://www.squadcast.com/blog/golden-signals-monitoring-from-first-principles`

- gRPC official website: `https://grpc.io`

- Service Mesh Interface: `https://smi-spec.io`

- Linkerd glossary and useful terms: `https://linkerd.io/service-mesh-glossary`

- Service meshes quick start and comparisons: `https://servicemesh.es`

- *Observability vs. Monitoring*: `https://dzone.com/articles/observability-vs-monitoring`

- *Monitoring and Observability — What's the Difference and Why Does It Matter?*: `https://thenewstack.io/monitoring-and-observability-whats-the-difference-and-why-does-it-matter`

- *The 4 Golden Signals of API Health and Performance in Cloud Native Applications*: `https://blog.netsil.com/the-4-golden-signals-of-api-health-and-performance-in-cloud-native-applications-a6e87526e74`

- Linkerd documentation: `https://linkerd.io/docs`

- *Service Mesh & Edge Computing Considerations*: `https://sunkur.medium.com/service-mesh-edge-computing-considerations-84126754d17a`

Edge Serverless and Event-Driven Architectures with Knative and Cloud Events

Serverless architecture reduces the costs of running distributed systems at scale. This use case is particularly useful in edge computing, where a lot of dedicated hardware and computational resources are used. This chapter covers how Knative can help you to implement APIs using serverless technologies. It also shows how to reduce costs and complexity using Knative for simple event-driven architectures and serverless functions to build your system. Across the chapter, we explain how Knative uses Cloud Events for its cloud event specification to call events, and how serverless can be helpful in the development of event-driven applications.

In this chapter, we're going to cover the following main topics:

- Serverless at the edge with Knative and Cloud Events

- Implementing serverless functions using Knative Serving

- Implementing a serverless API using traffic splitting with Knative

- Using declarative files in Knative

- Implementing events and event-driven pipelines using sequences with Knative Eventing

Technical requirements

For this chapter, you need the following:

- A single or multi-node K3s cluster using ARM devices with MetalLB installed and with the options to avoid Traefik being installed as the default ingress controller.

- kubectl configured to be used on your local machine to avoid using the `--kubeconfig` parameter.

- Clone the repository at `https://github.com/PacktPublishing/Edge-Computing-Systems-with-Kubernetes/tree/main/ch9` if you want to run the YAML configuration by using `kubectl apply` instead of copying the code from the book. Take a look at the code for Python and YAML configurations inside the `ch9` directory.

We are going to install Knative to implement simple use cases using serverless APIs and event-driven pipelines. Let's understand what serverless architectures are and how can they help in edge computing environments.

Serverless at the edge with Knative and Cloud Events

Edge computing is a paradigm that processes information near the source of data. This improves the response time of the application. It also saves bandwidth when the data is accessed because instead of getting data from the cloud, data is accessed near to the source. But one of the problems is that the services are always up and running. Here is where serverless can help to reduce costs, scaling down services when they are not used, helping to reduce additional costs compared with the traditional way of having services running all the time.

Ben Ellerby, in his Medium article called *Why Serverless will enable the Edge Computing Revolution*, mentions that *Serverless enables us to build applications and services without thinking about the underlying servers*. This refers to thinking more about the applications instead of managing infrastructure. In this way, serverless technologies and cloud services have been increasing in popularity in recent years. Serverless cloud services only charge you for the execution time when you are using the service. You can often find serverless services as small code functions. Serverless technologies enabled event-driven architectures to flourish, because of their simplicity and low cost to implement new functionalities. According to the `https://solace.com/` website, an event-driven architecture is a *software design pattern in which decoupled applications can asynchronously publish and subscribe to events via an event broker (modern messaging-oriented-middleware)*.

One of the key aspects to evaluate when building a new system is the cost of implementation. This will be a common scenario for choosing serverless technologies. Serverless technologies implemented in on-premises scenarios could take advantage of the temporal use of resources to execute serverless functions. Knative implements serverless functions and events that can be used to implement event-driven applications. In addition, an event specification such as Cloud Events can help to standardize the communication of your services and define events:

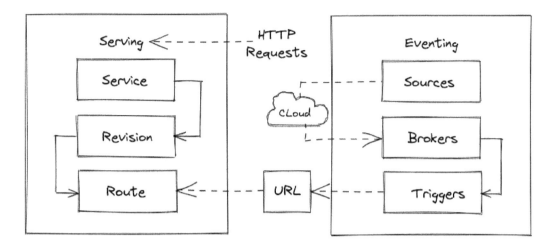

Figure 9.1 – Knative architecture

Knative was born in Google, and it was given to the community as an open source project. Knative consists of two parts: Serving and Eventing. With Knative Serving, you can create serverless functions in Kubernetes. Knative Serving implements the features of networking, autoscaling, and revision tracking. This abstraction gives the user the ability to focus more on the logic of the business instead of managing infrastructure. On the other hand, Knative Eventing gives the user the ability to implement event-driven architectures and call functions created with the Serving feature. You can configure your events to use different sources and broker types to manage your events depending on your use case. After choosing a source and broker that fit your scenario, you can trigger sequences or simple calls of your functions.

Cloud Events works together with Knative to give a standard structure to the events and have a uniform way to declare and call events. Cloud Events follows an event specification that is used to implement events. This structure has been adopted for different open source projects such as OpenFaaS, Tekton, Argo Events, Falco, Google Cloud Eventarc, and so on. The Cloud Events SDK is available for different programming languages such as Python and Go. This SDK will help you to describe cloud events through definitions such as ID, version of the cloud event specification, type, source, and content type.

Knative and Cloud Events provide a way to implement serverless functions and event-driven architectures at the edge, for low-resource devices, and a lightweight implementation that permits cost-saving in an edge computing scenario.

> **Important Note**
>
> For more information about Knative, you can visit its official documentation: `https://knative.dev/docs`. For Cloud Events, you can visit its official website: `https://cloudevents.io` or its specification 1.0, which is used in our examples: `https://github.com/cloudevents/spec/blob/v1.0.2/cloudevents/spec.md`.

Implementing serverless functions using Knative Serving

To start building our simple use cases for serverless and event-driven use cases, we have to install Knative with Serving, Eventing, channels, and brokers. In this case, we are going to use the basic options using in-memory channels and Knative Eventing Sugar Controller, which creates Knative resources based on labels in your cluster or namespace. So, let's start installing Knative Serving in the next section.

Installing Knative Serving

In this section, we are going to start installing Knative Serving, which will be used to implement serverless functions. Let's follow the next steps to install Knative Serving:

1. Install the Knative CLI with the following command:

   ```
   $ brew install kn
   ```

   To upgrade your current Knative binary, run the following:

   ```
   $ brew upgrade kn
   ```

2. Install the Knative Serving CRDs to install the serving components:

   ```
   $ kubectl apply -f https://github.com/knative/serving/
   releases/download/knative-v1.2.0/serving-crds.yaml
   $ kubectl apply -f https://github.com/knative/serving/
   releases/download/knative-v1.2.0/serving-core.yaml
   ```

   > **Note**
   >
   > To learn more about **Custom Resource Definitions** (**CRDs**) you can check out this link: https://docs.openshift.com/aro/3/dev_guide/creating_crd_objects.html. You can also check the CRD documentation from the Kubernetes official website with the next link: https://kubernetes.io/docs/concepts/extend-kubernetes/api-extension/custom-resources.

3. Now install the Contour ingress controller, which will be used as the default for Knative (this component is available for ARM):

   ```
   $ kubectl apply -f https://github.com/knative/
   net-contour/releases/download/knative-v1.2.0/contour.yaml
   ```

4. Install the network component of Knative for other functionalities using the previous ingress running the following command:

```
$ kubectl apply -f https://github.com/knative/
net-contour/releases/download/knative-v1.2.0/net-contour.
yaml
```

5. Then set Contour as the default ingress controller to be used by Knative:

```
$ kubectl patch configmap/config-network \
  --namespace knative-serving \
  --type merge \
  --patch '{"data":{"ingress-class":"contour.ingress.
networking.knative.dev"}}'
```

6. Get the IP that your Contour ingress controller created as the endpoint for your applications. In this case, we are going to call this IP EXTERNAL_IP:

```
$ EXTERNAL_IP="$(kubectl get svc envoy -n contour-
external  -o=jsonpath='{.status.loadBalancer.ingress[0].
ip}')"
```

7. Set the domain that Knative is going to use to expose your serverless applications:

```
$ KNATIVE_DOMAIN="$EXTERNAL_IP.nip.io"
$ kubectl patch configmap/config-domain \
--namespace knative-serving \
--type merge \
--patch '{"data":{"'$KNATIVE_DOMAIN'":""}}'
```

8. Now set the **Horizontal Pod Autoscaler** (**HPA**) feature of Knative Serving to run:

```
$ kubectl apply -f https://github.com/knative/serving/
releases/download/knative-v1.2.0/serving-hpa.yaml
```

9. Finally, perform simple troubleshooting for the Knative components running:

```
$ kubectl get pods -n knative-serving
```

This will return the state of the pods of your Knative Serving installation. These pods should have a ready status after a few minutes.

> **Important Note**
> To uninstall the components, you can use `kubectl delete` instead of `kubectl apply`.

Now Knative Serving is installed and ready to use. So, let's move on to create a simple serverless function using Knative Serving in the next section.

Creating a simple serverless function

Now it's time to use Knative Serving. In this section, we are going to run a sample API using Python and Flask. The code will look like this:

```
from flask import Flask
from flask import jsonify
import os
import socket
app = Flask(__name__)
host = socket.gethostname()
msg = os.environ['MESSAGE']
@app.route('/')
def index():
    return jsonify({"host":host,"msg":msg})
if __name__ == '__main__':
    app.run(host='0.0.0.0', port=5000, debug=True)
```

Every time you call the function, it is going to return the variable host with the container ID and msg with the value of the MESSAGE environment variable. This API will use port 5000. This Python program is already packaged in a container. It was built and published on Docker Hub as sergioarmgpl/app2demo.

> **Important Note**
>
> You can explore how to build and customize this code in the GitHub repository: https://github.com/sergioarmgpl/containers.

Now, to deploy this API as a serverless function using Knative, follow the next steps:

1. Create your function with the following command:

```
$ kn service create api \
--image sergioarmgpl/app2demo \
--port 5000 \
--env MESSAGE="Knative demo v1" \
--revision-name=v1
```

This command redirects port 5000 where your API is exposed in your container to the HTTP endpoint that Knative generates. It also receives the MESSAGE parameter with the Knative demo value and sets the revision of this function as v1. After running this command, you will get an output like this:

```
Service 'api' created to latest revision 'api-v1' is
available at URL:
http://api.default.192.168.0.54.nip.io
```

At the end of the output, you will find the endpoint for your function. In this output, we are assuming that the IP address assigned to the Contour ingress controller is 192.168.0.54, which is the same value assigned to the EXTERNAL_IP variable. Knative creates the necessary pods for this function in the default namespace. Refer to the *Installing Knative Serving* section for more information about to how to get the IP assigned to your Contour ingress.

2. Now, access your function using the EXTERNAL_IP variable defined in the *Installing Knative Serving* section, by running the following command:

```
$ curl http://api.default.$EXTERNAL_IP.nip.io
```

This command will return a JSON output in your terminal like this:

```
{
    "host": "api-v1-deployment-84f568857d-cxv9z",
    "msg": "Knative demo v1"
}
```

3. To monitor the pods created for your function, run this:

```
$ watch kubectl get pods
```

4. After 2 minutes of inactivity for your functions, the pods created to run your functions will be scaled down. If you execute watch kubectl get pods, you will see a similar output to this:

```
NAME            READY       STATUS
api-v1          2/2         Running
api-v1          2/2         Terminating
api-v1          1/2         Terminating
api-v1          0/2         Terminating
```

5. Open another terminal and execute `watch kubectl get pods`, and then call the function again. The pods of the function will be scaled up and you will see a similar output to this:

    ```
    NAME              READY    STATUS
    api-v1            0/2      Pending
    api-v1            0/2      ContainerCreating
    api-v1            1/2      Running
    api-v1            2/2      Running
    ```

    With the scaling to zero functionality, you can cut costs in your cloud infrastructure when your functions have an idle status after 2 minutes of inactivity.

    > **Important Note**
    >
    > The `watch` command might not be installed on your operating system. This can be installed with the `yum` or `apt` command on Linux, or the `brew` command on macOS.

6. Check the created services in the default namespace using the following command:

    ```
    $ kn service list
    ```

    Or, run the following command to check the available functions in a specific namespace:

    ```
    $ kn service list -n <YOUR_NAMESPACE>
    ```

7. To check your current revisions, run the following:

    ```
    $ kn revisions list
    ```

8. (*Optional*) If you don't want to create a public endpoint for your function, use the `--cluster-local` flag for the `kn` command to create a private endpoint. To create the same function but with a private endpoint, use the following command:

    ```
    $ kn service create api --cluster-local \
    --image sergioarmgp1/app2demo \
    --port 5000 \
    --env MESSAGE="Knative demo v1" \
    --revision-name=v1
    ```

    At the end of the output, you will see something like this:

    ```
    Service 'api' created to latest revision 'api-v1' is
    available at URL:
    http://api.default.svc.cluster.local
    ```

This endpoint will be the URL service that Knative creates for you, which is the same service object used in Kubernetes.

9. (*Optional*) To access this endpoint, you have to call it inside the cluster. To do this, create a client container that contains `curl`. Run the following command:

```
$ kubectl run curl -it --rm --image=curlimages/
curl:7.81.0 /bin/sh
```

Once the pod is created, you have to run the following command to access the function:

```
$ curl http://api.default.svc.cluster.local
```

The output will look like this:

```
{
  "host": "api-v1-deployment-776c896776-vxhhk",
  "msg": "Knative demo v1"
}
```

10. To delete the serverless function created in this section, run this:

```
$ kn service delete hello
```

Now, you know how to create serverless functions to implement a simple API using Knative Serving and scale to zero functionality to save costs. It's time to implement the traffic splitting functionality using Knative Serving in the next section.

Implementing a serverless API using traffic splitting with Knative

Knative has traffic splitting functionality that consists of distributing the traffic across two or more versions within a service but uses a proxy to implement this feature. By default, it uses Istio. For this implementation, we are using Contour, an Envoy-based proxy that consumes fewer resources than Istio. Both Istio and Contour use Envoy, a layer 7 proxy to implement service mesh capabilities such as traffic splitting. Traffic splitting could be used to implement deployment strategies such as canary and blue-green deployments, and also could be used to simulate faulty traffic for some basic chaos engineering scenarios. In this section, we are going to implement traffic splitting for the previous API function created in the *Creating a simple serverless function* section. In that section, we created a function called `api` with the revision name `v1`. Now we are going to update this function with another revision called `v2`. This revision just changes the `MESSAGE` value that is shown when you call the function. For this example, we are going to split traffic with 50% to revision `v1` and 50% to revision `v2`.

To implement this scenario, follow the next steps:

1. Update the current `api` function with the new revision, v2, which has the value of the `MESSAGE` variable with `Knative demo v2`, for this run:

```
$ kn service update api \
--env MESSAGE="Knative demo v2" \
--revision-name=v2
```

The output of this command will look like this:

```
Service hello created to latest revision 'api-v2' is
available at URL: http://api.default.192.168.0.54.nip.io
```

2. Let's check the revisions of our `api` function with the following command:

```
$ kn revisions list
```

With this command, you will see that 100% of the traffic will be processed by the v2 revision. The output will look like this:

```
NAME      SERVICE    TRAFFIC
api-v2    api        100%
api-v1    api
```

> **Important Note**
> We are omitting the TAGS, GENERATION, AGE, CONDITIONS, READY, and REASON fields of the output for learning purposes. We are assuming that the IP address assigned to the Contour ingress controller is 192.168.0.54, which is the same value assigned to the EXTERNAL_IP variable.

3. Set the traffic splitting to 50% for version v1 and 50% for version v2:

```
$ kn service update api \
--traffic api-v1=50 \
--traffic @latest=50
```

The expected output will look like this:

```
Service 'api' with latest revision 'api-v2' (unchanged)
is available at URL:
http://api.default.192.168.0.54.nip.io
```

You can also use `api-v2` instead of the `@latest` option. You can also customize your parameter with your own versions and different traffic splitting rates.

4. Let's check how traffic is distributed across the `api` function after setting the traffic splitting by running this:

```
$ kn revisions list
```

The output will look like this:

```
NAME      SERVICE   TRAFFIC
api-v2    api       50%
api-v1    api       50%
```

You will see that the traffic is split by 50% for each revision.

5. Let's send traffic to our function with a simple BASH loop script that you can stop with *Ctrl + C* by running the following command:

```
$ while true; do curl http://api.default.$EXTERNAL_
IP.nip.io;echo "";sleep 0.3; done
```

This command is going to continuously call your function that is split in to two versions every 0.3 seconds. The latest available revision will be running by default. In this case revision `v2` will be available for responses. After waiting a few seconds `v1` is provisioned and the output starts to show that the traffic is split by 50% for each revision. The output will look something like this:

```
{
    "host": "api-v1-deployment-85f6f977b5-hcgdz",
    "msg": "Knative demo v1"
}


{
    "host": "api-v1-deployment-85f6f977b5-hcgdz",
    "msg": "Knative demo v2"
}
```

Use *Ctrl + C* to stop the BASH loop.

6. If you want to check the pods of this traffic splitting, run the following command:

```
$ kubectl get pods -o=custom-columns=NAME:.metadata.
name,STATUS:.status.phase
```

The output will look like this:

```
NAME                                    STATUS
api-v1-deployment-85f6f977b5-jhss5      Running
api-v2-deployment-b97859489-mtvjm       Running
```

In this output, there are two pods running – one for revision `v1` and the other for `v2`. These pods are created on demand. By the default one of these revisions will be running if idle time was not exceeded to be called down. After requests start coming, the other revision is scaled up to start splitting the traffic between these pods by 50% each.

7. Finally, you can delete your API function with all your revisions running:

```
$ kn service delete api
```

Now you have learned how to use traffic splitting and revisions in Knative. Now let's go deep into Knative, learning how to use declarative files to create services in the next section.

Using declarative files in Knative

A good practice when creating environments is to create declarative definitions for your applications. Knative supports this with the `--target` flag. For example, if you want to change the previous example into a YAML file, you could use this flag. To do this, run the following command:

```
$ kn service create api --cluster-local \
--image sergioarmgpl/app2demo \
--port 5000 \
--env MESSAGE="Knative demo v1" \
--revision-name=v1 --target=api.yaml
```

This command outputs a YAML file with the definition of an API function, without a public endpoint. The output in the `api.yaml` file will look like this:

```
apiVersion: serving.knative.dev/v1
kind: Service
metadata:
  labels:
    networking.knative.dev/visibility: cluster-local
  name: api
  namespace: default
spec:
  template:
    metadata:
      annotations:
        autoscaling.knative.dev/max-scale: "5"
        containerConcurrency: 2
      name: api-v1
```

```
      spec:
        containers:
        - env:
          - name: MESSAGE
            value: "Knative demo v1"
          image: sergioarmgpl/app2demo
          name: ""
          ports:
          - containerPort: 5000
```

In the `annotations` section, you can configure different features that Knative provides; for example, autoscaling, rate limits, concurrency, and so on. In this case, we used `autoscaling.knative.dev/max-scale` to set the maximum replicas for the deployment of the function and `containerConcurrency` to set the number of simultaneous requests for each replica in the function.

Another example is how you can define the YAML for traffic splitting. Based on our previous traffic splitting example in the *Implementing a serverless API using traffic splitting with Knative* section, to generate the equivalent YAML configuration use the following command:

```
$ kn service update api \
--traffic api-v1=50 \
--traffic @latest=50 --target=api.yaml
```

The output will look like this:

```
apiVersion: serving.knative.dev/v1
kind: Service
metadata:
  labels:
    networking.knative.dev/visibility: cluster-local
  name: api
  namespace: default
spec:
  traffic:
  - latestRevision: true
    percent: 50
  - latestRevision: false
    percent: 50
    revisionName: api-v1
```

```
template:
  metadata:
    annotations:
      autoscaling.knative.dev/max-scale: "5"
      containerConcurrency: "2"
    name: api-v1
  spec:
    containers:
    - env:
      - name: MESSAGE
        value: "Knative demo v1"
      image: sergioarmgp1/app2demo
      name: ""
      ports:
      - containerPort: 5000
```

This is a desirable feature and best practice. To have declarative definitions for creating your functions and other Knative objects, you can explore the official documentation of Knative to find examples of declarative definitions. Now it's time to move on to install another feature, Knative Eventing, in the next section.

Implementing events and event-driven pipelines using sequences with Knative Eventing

Knative provides Eventing components to implement event-driven architectures. We are going to explore a simple Eventing pipeline with Knative using the lightweight in-memory channel component to implement two simple events that call a service showing a message. In the second part, we are going to implement a simple sequence that calls two servers sequentially, one after the other, showing custom messages. So, let's get started with the first part to implement simple events.

Installing Knative Eventing

Before creating our events, we need to install all the Knative components. We are going to use the in-memory channel to manage our events, which is the simplest and most lightweight channel implemented in Knative, and Sugar Controller to provision Knative Eventing resources in namespaces using labels. To get started with installing Knative Eventing, follow the next steps:

1. Install the Knative Eventing CRDs:

    ```
    $ kubectl apply -f https://github.com/knative/eventing/
    releases/download/knative-v1.2.0/eventing-crds.yaml
    ```

2. Install Knative Eventing core components by running the following command:

```
$ kubectl apply -f https://github.com/knative/eventing/
releases/download/knative-v1.2.0/eventing-core.yaml
```

3. Now install the in-memory channel component by running this:

```
$ kubectl apply -f https://github.com/knative/eventing/
releases/download/knative-v1.2.0/in-memory-channel.yaml
```

4. Now install the MT channel broker, which is a lightweight and simple implementation to use the in-memory channel:

```
$ kubectl apply -f https://github.com/knative/eventing/
releases/download/knative-v1.2.0/mt-channel-broker.yaml
```

5. Finally, install Knative Eventing Sugar Controller, which reacts to special labels and annotations and produces Eventing resources:

```
$ kubectl apply -f https://github.com/knative/eventing/
releases/download/knative-v1.2.0/eventing-sugar-
controller.yaml
```

6. Check whether all the components have a READY status by running the following command:

```
$ kubectl get pods -n knative-eventing -o=custom-
columns=NAME:.metadata.name,STATUS:.status.phase
```

You will see a similar output to this:

```
NAME                                   STATUS
mt-broker-filter-574dc4457f-pjs7z      Running
imc-dispatcher-7fcb4b5d8c-qxrq2        Running
mt-broker-controller-8d979648f-6st56   Running
sugar-controller-6dd4c4bc5f-76kqc      Running
mt-broker-ingress-5ddd6f8b5d-h94z5     Running
eventing-webhook-5968f79978-5nhlc      Running
eventing-controller-58875c5478-n8xzl   Running
imc-controller-86cd7b7857-hpcpq        Running
```

Now you have installed all the necessary components to implement a simple event-driven pipeline using Knative. Let's move to the next section to learn how to implement events.

Implementing a simple event

Now it's time to implement some basic events. This scenario consists of creating two services and calling them with their attribute type. First, let's explore the code inside the container that is in Docker Hub called `sergioarmgpl/app3demo`. The code used is this:

```
from flask import Flask, request
from cloudevents.http import from_http
app = Flask(__name__)

@app.route("/", methods=["POST"])
def route():
    event = from_http(request.headers, request.get_data())
    app.logger.warning(event)
    return "", 204

if __name__ == "__main__":
    app.run(debug=True, host='0.0.0.0',port=5000)
```

This code receives the call and transforms the data of the requests using the Cloud Events library to output the event with the `app.logger.warning` function implemented in Flask. So, every time the application is called in the / route path, it is going to show the information of the request that is calling the container using the Cloud Events structure format in the logs. In this case, we are not returning any data in response. It just returns HTTP status response code `204`, which refers to a successful request call. You can also customize this code if necessary to fit your needs.

Now we have to create two services using YAML definitions. The first service will be called `api-demo`, and the second `api-demo2`. These services will be called every time the broker is called, sending their cloud event's attributes. When the attribute type is set to `event.show`, the `api-demo` service is called, and when the broker is called with the `attribute` type set to `event.show.2`, the `api-demo2` service will be called. Both services are configured to listen on port `5000` and forward requests to port `80` to properly work with Knative Eventing.

To start implementing the first scenario, follow the next steps:

1. Create and inject the `event-demo` namespace where the event is going to be created:

    ```
    $ cat <<EOF | kubectl apply -f -
    apiVersion: v1
    kind: Namespace
    metadata:
      name: event-demo
    ```

```
      labels:
          eventing.knative.dev/injection: enabled
    EOF
```

2. Create the default broker to use for this implementation:

```
$ cat <<EOF | kubectl apply -f -
apiVersion: eventing.knative.dev/v1
kind: Broker
metadata:
  name: default
  namespace: event-demo
  annotations:
    eventing.knative.dev/broker.class:
MTChannelBasedBroker
EOF
```

3. Deploy the container that is going to process the event:

```
$ cat <<EOF | kubectl apply -f -
apiVersion: apps/v1
kind: Deployment
metadata:
  labels:
    app: api-demo
  name: api-demo
  namespace: event-demo
spec:
  replicas: 1
  selector:
    matchLabels:
      app: api-demo
  template:
    metadata:
      labels:
        app: api-demo
    spec:
      containers:
```

```
          - image: sergioarmgp1/app3demo
            name: app3demo
            imagePullPolicy: Always
    EOF
```

4. Create the service for this api-demo deployment:

```
$ cat <<EOF | kubectl apply -f -
apiVersion: v1
kind: Service
metadata:
  labels:
    app: api-demo
  name: api-demo
  namespace: event-demo
spec:
  ports:
  - port: 80
    protocol: TCP
    targetPort: 5000
  selector:
    app: api-demo
  type: ClusterIP
EOF
```

5. Create a trigger to be consumed by the service:

```
$ cat <<EOF | kubectl apply -f -
apiVersion: eventing.knative.dev/v1
kind: Trigger
metadata:
  name: api-demo
  namespace: event-demo
spec:
  broker: default
  filter:
    attributes:
      type: event.show
```

```
    subscriber:
      ref:
        apiVersion: v1
        kind: Service
        name: api-demo
  EOF
```

6. Create a pod in the event-demo namespace to call the broker. This broker is going to call our pod that shows the message **Simple Event using Knative**. To create this pod, run this:

```
$ kubectl run -n event-demo curl -it --rm
--image=curlimages/curl:7.81.0 /bin/sh
```

7. Inside this pod, run the curl command to send a request to the broker. The broker will take the parameters of the previously implemented cloud event to send it to your pod. To call the broker, run this:

```
$ curl -v "broker-ingress.knative-eventing.svc.cluster.
local/event-demo/default" \
-X POST \
-H "Ce-Id: call-api-demo" \
-H "Ce-specversion: 1.0" \
-H "Ce-Type: event.show" \
-H "Ce-Source: test-send" \
-H "Content-Type: application/json" \
-d '{"msg":"Simple Event using Knative."}'
```

The output will look like this:

```
* Connected to broker-ingress.knative-eventing.svc.
cluster.local (10.43.130.39) port 80 (#0)
> POST /event-demo/default HTTP/1.1
> Host: broker-ingress.knative-eventing.svc.cluster.local
> User-Agent: curl/7.81.0-DEV
> Accept: */*
> Ce-Id: 536808d3-88be-4077-9d7a-a3f162705f79
> Ce-specversion: 0.3
> Ce-Type: dev.knative.myevents.api-demo
> Ce-Source: dev.knative.myevents/api-demo-source
> Content-Type: application/json
> Content-Length: 37
```

```
>
* Mark bundle as not supporting multiuse
< HTTP/1.1 202 Accepted
< Allow: POST, OPTIONS
< Date: Thu, 24 Feb 2022 05:30:13 GMT
< Content-Length: 0
<
* Connection #0 to host broker-ingress.knative-eventing.
svc.cluster.local left intact
```

8.   To exit, run the next command inside the pod:

```
$ exit
```

9.   Now inspect the logs of the pod by running the following command:

```
$ kubectl -n event-demo logs -l app=api-demo --tail=50
```

Or, if you want to see the log in real time, when you call the broker that calls your pod, run the following command:

```
$ kubectl -n event-demo logs -f -l app=api-demo
```

The output will look like this:

```
* Serving Flask app 'index' (lazy loading)
* Environment: production
   WARNING: This is a development server. Do not use it
in a production deployment.
   Use a production WSGI server instead.
* Debug mode: on
* Running on all addresses.
   WARNING: This is a development server. Do not use it
in a production deployment.
* Running on http://10.42.0.42:5000/ (Press CTRL+C to
quit)
* Restarting with stat
* Debugger is active!
* Debugger PIN: 110-221-376
[2022-02-27 06:02:02,107] WARNING in index:
{'attributes': {'specversion': '1.0', 'id': 'call-
api-demo', 'source': 'test-send', 'type': 'event.
show', 'datacontenttype': 'application/json',
```

```
'knativearrivaltime': '2022-02-27T06:02:02.069191004Z',
'time': '2022-02-27T06:02:02.107288+00:00'}, 'data':
{'msg': 'Simple Event using Knative.'}}
```

As you can see, the pod got the `msg` value `Simple Event using Knative.` and it's printed in the logs of the pod. This means that when you call the broker, the trigger calls the pod exposed using the service that was previously created.

Let's say, for example, that you want to create another event, using the same image. This time, let's call it `api-demo2` for the second service. Create the next YAML definitions:

1. To create the `api-demo2` deployment, run the following:

```
$ cat <<EOF | kubectl apply -f -
apiVersion: apps/v1
kind: Deployment
metadata:
  labels:
    app: api-demo2
  name: api-demo2
  namespace: event-demo
spec:
  replicas: 1
  selector:
    matchLabels:
      app: api-demo2
  template:
    metadata:
      labels:
        app: api-demo2
    spec:
      containers:
      - image: sergioarmgp1/app3demo
        name: app4
        imagePullPolicy: Always
EOF
```

2. Create the service for this `api-demo2` deployment:

```
$ cat <<EOF | kubectl apply -f -
apiVersion: v1
```

```
kind: Service
metadata:
  labels:
    app: api-demo2
  name: api-demo2
  namespace: event-demo
spec:
  ports:
  - port: 80
    protocol: TCP
    targetPort: 5000
  selector:
    app: api-demo2
  type: ClusterIP
EOF
```

3. Create a trigger that launches api-demo2, and let's call the attribute type event.show.2 to call the api-demo2 service, which points to the api-demo2 deployment:

```
$ cat <<EOF | kubectl apply -f -
apiVersion: eventing.knative.dev/v1
kind: Trigger
metadata:
  name: api-demo2
  namespace: event-demo
spec:
  broker: default
  filter:
    attributes:
      type: event.show.2
  subscriber:
    ref:
      apiVersion: v1
      kind: Service
      name: api-demo2
EOF
```

4. In the previously created `curl` pod, run the following command:

```
$ curl -v "broker-ingress.knative-eventing.svc.cluster.
local/event-demo/default" \
-X POST \
-H "Ce-Id: call-api-demo2" \
-H "Ce-specversion: 1.0" \
-H "Ce-Type: event.show.2" \
-H "Ce-Source: test-send" \
-H "Content-Type: application/json" \
-d '{"msg":"Simple Event using Knative."}'
```

5. Check the logs in the new `api-demo2` deployment with the following command:

```
$ kubectl -n event-demo logs -l app=api-demo2 --tail=50
```

6. The log will look like this:

```
* Serving Flask app 'index' (lazy loading)
* Environment: production
  WARNING: This is a development server. Do not use it
in a production deployment.
  Use a production WSGI server instead.
* Debug mode: on
* Running on all addresses.
  WARNING: This is a development server. Do not use it
in a production deployment.
* Running on http://10.42.0.43:5000/ (Press CTRL+C to
quit)
* Restarting with stat
* Debugger is active!
* Debugger PIN: 602-982-734
[2022-02-27 06:16:07,689] WARNING in index:
{'attributes': {'specversion': '1.0', 'id': 'call-
api-demo2', 'source': 'test-send', 'type': 'event.
show.2', 'datacontenttype': 'application/json',
'knativearrivaltime': '2022-02-27T06:16:07.654229185Z',
'time': '2022-02-27T06:16:07.688895+00:00'}, 'data':
{'msg': 'Simple Event using Knative2.'}}
```

Now you have created two basic events using Knative Eventing. This can help you to implement simple and lightweight event-driven architectures. Now, it's time to explore how to use the Sequence feature of Knative Eventing to create and run simple pipelines using an event-driven architecture.

Using sequences to implement event-driven pipelines

Another common use case for event-driven architectures is to trigger a series of steps one after the other to automate workflows. In those cases, you can use the Sequence object of Knative. In this example, we are going to create a sequence that consists of two steps. Each step prints the MESSAGE variable, which contains the number of the step that is running. This sequence is going to be called using a trigger. We are going to call the trigger using the curl command. This is a simple example pipeline using event-driven architecture. Let's get started by following the next steps:

1. Create the sequence-demo namespace with the eventing.knative.dev/injection: enabled label. When Knative Eventing detects this label in your namespace, it is going to create the default Knative broker. This is possible thanks to the Knative Sugar Controller previously installed. So, let's create the namespace by running the following command:

```
$ cat <<EOF | kubectl apply -f -
apiVersion: v1
kind: Namespace
metadata:
  name: sequence-demo
  labels:
      eventing.knative.dev/injection: enabled
EOF
```

2. Create step1 using the Knative Service definition file by running the following command:

```
$ cat <<EOF | kubectl apply -f -
apiVersion: serving.knative.dev/v1
kind: Service
metadata:
  name: step1
  namespace: sequence-demo
spec:
  template:
    spec:
      containers:
        - image: sergioarmgpl/app4demo
```

```
        ports:
         - containerPort: 5000
        env:
          - name: MESSAGE
            value: "step1"
   EOF
```

3. Now create `step2` by running the following:

```
$ cat <<EOF | kubectl apply -f -
apiVersion: serving.knative.dev/v1
kind: Service
metadata:
  name: step2
  namespace: sequence-demo
spec:
  template:
    spec:
      containers:
        - image: sergioarmgp1/app4demo
          ports:
           - containerPort: 5000
          env:
            - name: MESSAGE
              value: "step2"
   EOF
```

> **Important Note**
>
> We are using the `containerPort` parameter in the service definition, to define a custom port where our container is listening, in order to talk with Knative Eventing. By default, Knative uses port `80` to listen to services.

4. Let's create our sequence object called `sequence-demo` to run the steps as a small pipeline using the in-memory channel for messaging:

```
$ cat <<EOF | kubectl apply -f -
apiVersion: flows.knative.dev/v1
kind: Sequence
```

```
metadata:
  name: sequence
  namespace: sequence-demo
spec:
  channelTemplate:
    apiVersion: messaging.knative.dev/v1
    kind: InMemoryChannel
  steps:
    - ref:
        apiVersion: serving.knative.dev/v1
        kind: Service
        name: step1
    - ref:
        apiVersion: serving.knative.dev/v1
        kind: Service
        name: step2
EOF
```

5. Create the trigger that we are going to use. We are going to define an attribute to call it. In this case, every time we call an event with the `type` attribute with the value `event.call.sequence`, it is going to call our sequence:

```
$ cat <<EOF | kubectl apply -f -
apiVersion: eventing.knative.dev/v1
kind: Trigger
metadata:
  name: sequence-trigger
  namespace: sequence-demo
spec:
  broker: default
  filter:
    attributes:
      type: event.call.sequence
  subscriber:
    ref:
      apiVersion: flows.knative.dev/v1
      kind: Sequence
```

```
      name: sequence
EOF
```

6. Now let's create a `curl` pod inside the `sequence-demo` namespace to call our sequence using the endpoint of our broker:

```
$ kubectl run -n sequence-demo curl -it --rm
--image=curlimages/curl:7.81.0 /bin/sh
```

7. Inside the pod, run the following `curl` command:

```
$ curl -v "broker-ingress.knative-eventing.svc.cluster.
local/sequence-demo/default" \
-X POST \
-H "Ce-Id: call-sequence-demo" \
-H "Ce-specversion: 1.0" \
-H "Ce-Type: event.call.sequence" \
-H "Ce-Source: test-sequence" \
-H "Content-Type: application/json" \
-d '{"SOME_VARIABLE":"Simple Sequence using Knative."}'
```

This is going to show an output like this:

```
*     Trying 10.43.130.39:80...
* Connected to broker-ingress.knative-eventing.svc.
cluster.local (10.43.130.39) port 80 (#0)
> POST /sequence-demo/default HTTP/1.1
> Host: broker-ingress.knative-eventing.svc.cluster.local
> User-Agent: curl/7.81.0-DEV
> Accept: */*
> Ce-Id: call-sequence-demo
> Ce-specversion: 1.0
> Ce-Type: event.call.sequense
> Ce-Source: test-sequence
> Content-Type: application/json
>
* Mark bundle as not supporting multiuse
< HTTP/1.1 202 Accepted
< Allow: POST, OPTIONS
< Date: Mon, 28 Feb 2022 01:00:50 GMT
```

```
< Content-Length: 0
<
* Connection #0 to host broker-ingress.knative-eventing.
svc.cluster.local left intact
```

8. Exit the pod and check the output of the pod for *step 1* by running the following:

```
$ kubectl logs deploy/step1-00001-deployment -n sequence-
demo user-container
```

You will see an output like this:

```
[2022-02-28 01:06:54,364] WARNING in index: b'{"SOME_
VARIABLE":"Simple Sequence using Knative."}'
[2022-02-28 01:06:54,365] WARNING in index: step1
```

This is going to receive the SOME_VARIABLE variable, sent by the curl command, which could be used to customize your sequence.

9. Now check the output for *step 2* by running the following:

```
$ kubectl logs deploy/step2-00001-deployment -n sequence-
demo user-container
The output will look like:
[2022-02-28 01:07:02,623] WARNING in index: b'{\n   "ENV_
VAR": "step1"\n}\n'
[2022-02-28 01:07:02,624] WARNING in index: step2
```

This is going to show the ENV_VAR value sent by the previous step and the current environment variable showing the step currently running – in this case, *step 2*.

10. After a few minutes of being idle, the deployments for the steps in the namespace will scale down and will scale up every time you call it.

We have finished with the basics of serverless and event-driven pipelines using Knative. It's time to finish this chapter.

Summary

In this chapter, we learned how to implement public serverless and internal serverless functions using Knative Serving and use the features of traffic splitting. We also learned how to implement simple events and a sequence of events to implement small event-driven architectures using Knative Eventing, and how to integrate and standardize API event calls using the Cloud Events Python SDK. In the next chapter, we are going to learn how to use databases at the edge to add more functionality to edge systems using K3s.

Questions

Here are a few questions to validate your new knowledge:

1. What are the use cases for serverless architectures?

2. What is a serverless function?

3. What are the advantages of serverless technology?

4. How can I implement a serverless function using Knative?

5. How can I implement an event using Knative?

6. How can I implement an event-driven pipeline using Knative?

7. How does Cloud Events help you to implement events?

Further reading

You can refer to the following references for more information on the topics covered in this chapter:

- *Why Serverless will enable the Edge Computing Revolution*: https://medium.com/serverless-transformation/why-serverless-will-enable-the-edge-computing-revolution-4f52f3f8a7b0

- *What is edge serverless*: https://www.stackpath.com/edge-academy/what-is-edge-serverless

- *AI/ML, edge and serverless computing top priority list for the year ahead*: https://www.redhat.com/en/blog/aiml-edge-and-serverless-computing-top-priority-list

- Running Knative on Raspberry Pi: https://github.com/csantanapr/knative-pi

- Install Knative Serving using YAML: https://knative.dev/docs/install/serving/install-serving-with-yaml/#install-a-networking-layer

- Cloud Events website: https://cloudevents.io

- Cloud Events SDK: https://github.com/cloudevents/sdk-python

- CloudEvents – version 1.0.2: https://github.com/cloudevents/spec/blob/v1.0.2/cloudevents/spec.md

- A Hello World Python example with Knative Eventing: https://github.com/knative/docs/tree/main/code-samples/eventing/helloworld/helloworld-python

- Sending events to the broker: https://knative.dev/docs/eventing/getting-started/#sending-events-to-the-broker

- Using Sequence with Broker and Trigger: https://knative.dev/docs/eventing/flows/sequence/sequence-with-broker-trigger

10
SQL and NoSQL Databases at the Edge

When you have to create an edge system, a critical task is storing your data. For this, you have to take into consideration the resources that you have, the processor that your devices are using, and the type of data that you want to store. CAP theorem states that distributed data stores only provide two of the following guarantees: consistency, availability, and partition tolerance. So, this theorem can help you to decide which type of database is best according to your system needs. In this chapter, we are going to learn how to deploy different database types to run on edge systems using K3s and ARM devices. These examples include different techniques such as using ConfigMaps and Secrets to deploy your databases.

In this chapter, we're going to cover the following main topics:

- CAP theorem for SQL and NoSQL databases
- Creating a volume to persist your data
- Using MySQL and MariaDB SQL databases
- Using a Redis key-value NoSQL database
- Using a MongoDB document-oriented NoSQL database
- Using a PostgreSQL object-relational SQL database
- Using a Neo4j graph NoSQL database

Technical requirements

To deploy the databases in this chapter, you need the following:

- A single- or multi-node K3s cluster using ARM devices with MetalLB, and Longhorn storage installed. If you are using Raspberry Pi devices, you will need at least 4 GB of RAM and at least the 4B model. Each node must have the Ubuntu ARM64 operating system in order to support the ARMv8 architecture, necessary for some deployments in this chapter.

- `kubectl` configured to be used on your local machine, to avoid using the `--kubeconfig` parameter.

- Clone the repository at `https://github.com/PacktPublishing/Edge-Computing-Systems-with-Kubernetes/tree/main/ch10` if you want to run the YAML configuration by using `kubectl apply` instead of copying the code from the book. Take a look at the directory `yaml` for the YAML examples inside the `ch10` directory.

With this, you can deploy the databases explained in this chapter. So, let's get started learning about CAP theorem first, to choose the right database for your specific use case.

CAP theorem for SQL and NoSQL databases

CAP theorem was defined by Eric Brewer in 1999 and presented at the 19th Annual ACM Symposium on **Principles of Distributed Computing (PODC)** in 2000. This theorem states that a distributed data store can only provide two of the following guarantees:

- **Consistency**: This means when reading information, the data store returns the most recent written data or returns an error if it fails. This refers to regular SQL databases that use atomic operations to guarantee that data is written. If not, the system automatically rolls back to a previous data state.

- **Availability**: This means that all reads contain data, but it might not be the most recent.

- **Partition tolerance**: This is the most desired feature in a distributed system. It means that data is distributed in several nodes, helping to reduce downtime for the database. This means that if a node is down, just a small portion of data will be inaccessible:

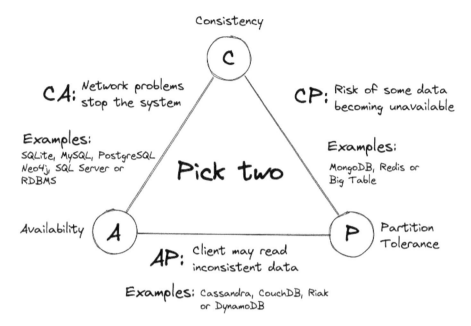

Figure 10.1 – CAP theorem diagram

This theorem is commonly used as a point of reference to design strong distributed systems in the context of data. Now, in the CAP theorem diagram (*Figure 10.1*), represented as a triangle, you can see the different sides, and how each side has a relationship with the other sides. Let's explore these sides and give examples of databases:

- **CA**: On this side, we can classify databases that have consistency and availability. Here, we can find SQLite, which is a very simple database. MySQL and PostgreSQL are very popular open source databases. SQL Server is a proprietary database from Microsoft and Neo4j is a graph database. Each of these databases tries to guarantee consistency and availability. These guarantees can be found on **relational database management system** (**RDBMS**)-based databases. But as we have mentioned, Neo4j is classified on this side of the triangle. Something important on this side is that the database will fail if the network is down.

- **CP**: On this side, you can find databases that provide consistency and partition tolerance. This means that databases such as Mongo and Redis use algorithms to write information that guarantee the consistency of data. For example, MongoDB uses the reader-writer algorithm to write in the database. Redis uses a similar algorithm to write data. Talking about partition tolerance, MongoDB can distribute information across nodes. This gives MongoDB the ability to partition the data. This is sharding, which provides the partition tolerance feature to MongoDB. Other databases based on Bigtable work similarly. Those Bigtable-based databases usually read data from distributed buckets of information across the cloud. The problem in CP is the risk that some data will become unavailable when a node or source of data is down.

- **AP**: On this side, databases look for availability and partition tolerance. Here, we can find databases such as Cassandra, CouchDB, Riak, DynamoDB, and databases based on Cassandra. For example, Cassandra has high availability, using its masterless technique to scale servers, but it doesn't guarantee the consistency of data. That's a common issue in some NoSQL databases.

Before deciding which database is right for you, let's explore what a relational and no-relational database is. A **relational database** is a database where the data is structured. This means that data is organized in tables, rows, and columns. These tables have relationships and dependencies. These databases use **Structured Query Language** (**SQL**) to manage the information. Relational databases are also called **SQL databases**. They also use ACID operations. This stands for atomicity, consistency, isolation, and durability; these properties in data guarantee data integrity when errors and failures happen. Some examples are MySQL, PostgreSQL, and SQL Server.

A **non-relational database** is not structured. It doesn't use a table, row, and column data schema. Instead, it uses a storage model optimized for specific requirements of the type of data being stored. Some of these types of data could be JSON documents, key values, and so on. These databases are also called **NoSQL databases**. These databases don't use ACID operations. They look for availability and partition tolerance for data. Some examples are MongoDB, Redis, Neo4j, and Cassandra. When choosing the right database, you can evaluate some of these questions:

- Which of the guarantees of consistency, partition tolerance, and availability does my system need? According to this, which database fits my system needs best?

- Does my database need to support the SQL language to query information?

- Do I need a database that supports the SQL language?

- Is my data not structured as JSON documents or do I need something structured as tables?

- What type of data am I storing? Do I need a SQL or NoSQL database?

- Do I need consistency, availability, or partition tolerance? Which of these components is important for my system?

- How many resources is my database going to use? How many simultaneous connections is my system expected to handle?

- Do I need to replicate information, implement rate limits, or have any other specific features in my database?

- How fast is my database at writing and reading data?

- How can I do replication or scaling on my database?

These and other questions could be important when choosing the right database. So, this chapter focuses on giving you a quick start when choosing the right database, using CAP theorem and some examples of how to deploy some SQL and NoSQL databases mentioned in the CAP theorem description. These SQL and NoSQL databases will be deployed at the edge in a K3s cluster using containers.

> **Important Note**
>
> You can find in the *Further reading* section some links to learn more about SQL and NoSQL databases, the official web links for the databases explained in this chapter, and complementary links to evaluate which database is best for your use case. A complementary theorem that you can use is the PACELC theorem. This looks for the trade-offs between latency and consistency when data is replicated.

Now let's move on to create a volume to persist your data before performing the deployment of your selected database.

Creating a volume to persist your data

Before we start deploying our databases, let's create a volume to store data first. For this, we have two options. One is to use a directory inside the server. This means that in order to not lose data, your Pods have to be provisioned in the same node as where your volume was created the first time. If you don't want to depend on which node your pods are running, you have to choose a second option, which is to use a storage driver. If that's your case, we are going to use Longhorn as our storage driver option. Now, let's create our storage first, using a local directory. For this, follow the next steps:

1. Create a **PersistentVolume** using the /mnt/data directory in the node to store data:

```
$ cat <<EOF | kubectl apply -f -
apiVersion: v1
kind: PersistentVolume
metadata:
  name: db-pv-volume
  labels:
    type: local
spec:
  storageClassName: manual
  capacity:
    storage: 5Gi
  accessModes:
    - ReadWriteOnce
  hostPath:
    path: "/mnt/data"
EOF
```

2. Create a **PersistentVolumeClaim** using 5 GB of storage:

```
$ cat <<EOF | kubectl apply -f -
apiVersion: v1
kind: PersistentVolumeClaim
metadata:
  name: db-pv-claim
spec:
  storageClassName: manual
  accessModes:
    - ReadWriteOnce
  resources:
    requests:
      storage: 5Gi
EOF
```

If you want to use Longhorn as your storage, follow the next steps:

1. Create a PersistentVolumeClaim with 5 GB of storage, this time, using Longhorn:

```
$ cat <<EOF | kubectl apply -f -
apiVersion: v1
kind: PersistentVolumeClaim
metadata:
  name: db-pv-claim
spec:
  accessModes:
    - ReadWriteOnce
  storageClassName: longhorn
  resources:
    requests:
      storage: 5Gi
EOF
```

This is a critical step to persist and avoid losing your data. In the next sections, we are going to start deploying our databases, starting with basic configuration and adding more complex configurations such as using **ConfigMaps** and **Secrets** to perform more production-ready deployments. But first, let's start with MySQL and MariaDB, very popular databases used across the internet.

Using MySQL and MariaDB SQL databases

MySQL is a relational database that uses the SQL language to read and write information. It's one of the most used databases on the internet. MariaDB is a fork of MySQL and the version used in this example is fully compatible with MySQL. It's a very fast SQL database and it's simple to use. After this brief introduction to MySQL, let's get started deploying this database by following the next steps:

1. Create the MySQL deployment creating a PersistentVolumeClaim called `db-pv-claim`:

    ```
    $ cat <<EOF | kubectl apply -f -
    apiVersion: apps/v1
    kind: Deployment
    metadata:
      name: mysql
    spec:
      selector:
        matchLabels:
          app: mysql
      strategy:
        type: Recreate
      template:
        metadata:
          labels:
            app: mysql
        spec:
          containers:
          - image: mysql:8.0.28-oracle
            name: mysql
            env:
            - name: MYSQL_ROOT_PASSWORD
              value: password
            ports:
            - containerPort: 3306
              name: mysql
            volumeMounts:
            - name: mysql-persistent-storage
              mountPath: /var/lib/mysql
          volumes:
    ```

```
       - name: mysql-persistent-storage
         persistentVolumeClaim:
           claimName: db-pv-claim
EOF
```

> **Important Note**
>
> Instead of using MySQL, you can use MariaDB, which is fully compatible with MySQL version 5.6. To do this, change the `mysql:8.0.28-oracle` image to `arm64v8/mariadb:latest` and the `MYSQL_ROOT_PASSWORD` variable to `MARIADB_ROOT_PASSWORD`. You can also check for other image versions in `https://hub.docker.com` for MySQL and MariaDB images. For this deployment, the password is `password`. The images used for the deployment are both designed to run on ARM devices. In the case of MySQL reinstallation using local storage, you have to delete the content inside the `/mnt/data` directory using the `rm -R /mnt/data` command to avoid errors.

2. Now let's create our service to access MySQL using a service:

```
$ cat <<EOF | kubectl apply -f -
apiVersion: v1
kind: Service
metadata:
  name: mysql
spec:
  ports:
  - port: 3306
  selector:
    app: mysql
  clusterIP: None
EOF
```

3. To test whether your MySQL deployment works, you can access your deployment pod by running this:

```
$ kubectl exec -it $(kubectl get pods -l app=mysql
--output=jsonpath={..metadata.name}) -- bash
```

Inside the pod, run the following command to connect to your database:

```
$ mysql -h localhost -uroot -ppassword
```

Now, the prompt will change to mysql>. Let's create a simple database, EXAMPLE, with the VALUE_TABLE table, and insert and list some records. To do this, run the following commands and you will see output like this:

```
mysql> CREATE DATABASE EXAMPLE;
Query OK, 1 row affected (0.02 sec)


mysql> USE EXAMPLE;
Database changed


mysql> CREATE TABLE VALUE_TABLE (ID INT PRIMARY KEY NOT
NULL,VALUE INT NOT NULL);
Query OK, 0 rows affected (0.10 sec)


mysql> INSERT INTO VALUE_TABLE (ID,VALUE) VALUES (1,123);
Query OK, 1 row affected (0.03 sec)


mysql> SELECT * FROM VALUE_TABLE;
+----+-------+
| ID | VALUE |
+----+-------+
|  1 |   123 |
+----+-------+
1 row in set (0.00 sec)
```

4. Finally, delete the table and database with the following commands:

```
mysql> DROP TABLE VALUE_TABLE;
Query OK, 0 rows affected (0.07 sec)


mysql> DROP DATABASE EXAMPLE;
Query OK, 0 rows affected (0.05 sec)


mysql> EXIT
Bye
```

Now you have learned how to use MySQL with this basic deployment and example. Now let's move on to learn how Redis works.

Using a Redis key-value NoSQL database

Now it's time to use Redis as our key-value database. Redis is a nice key-value database that doesn't consume many resources. All its data is stored in memory. It has very interesting types of data such as hash keys, lists, and sets. It also implements publisher-subscriber and streaming features to implement channels of communication and simple broker functionalities. For our Redis deployment, we are going to use a custom configuration to set the password for Redis, and a storage volume to prevent losing data. To use Redis in your cluster, follow the next steps:

1. Create the **ConfigMap** to use a custom configuration with the password K3s123- and the / data directory to store Redis data:

    ```
    $ cat <<EOF | kubectl apply -f -
    apiVersion: v1
    kind: ConfigMap
    metadata:
      name: redis-configmap
    data:
      redis-config: |
        dir /data
        requirepass YOUR_PASSWORD
    EOF
    ```

2. Create the deployment for Redis using the previous ConfigMap called redis-configmap and mounted as the redis.conf file. We also use the PersistentVolumeClaim called db-pv-claim, and some resource limits for the deployment, setting the CPU and memory. Let's create the deployment by running the following command:

    ```
    $ cat <<EOF | kubectl apply -f -
    apiVersion: apps/v1
    kind: Deployment
    metadata:
      labels:
        run: redis
      name: redis
    spec:
      replicas: 1
      selector:
        matchLabels:
          run: redis
    ```

```
template:
  metadata:
    labels:
      run: redis
  spec:
    containers:
    - name: redis
      image: arm64v8/redis:6.2
      command:
        - redis-server
        - /redisconf/redis.conf
      ports:
      - containerPort: 6379
      resources:
        limits:
          cpu: "0.2"
          memory: "128Mi"
      volumeMounts:
      - mountPath: "/data"
        name: redis-storage
      - mountPath: /redisconf
        name: config
    volumes:
      - name: config
        configMap:
          name: redis-configmap
          items:
          - key: redis-config
            path: redis.conf
      - name: redis-storage
        persistentVolumeClaim:
          claimName: db-pv-claim
EOF
```

3. Now create the `redis`, which points to port `6379` in our `redis` deployment:

```
$ cat <<EOF | kubectl apply -f -
apiVersion: v1
kind: Service
metadata:
  labels:
    run: redis
  name: redis
spec:
  ports:
  - port: 6379
    protocol: TCP
    targetPort: 6379
  selector:
    run: redis
  type: ClusterIP
EOF
```

This service creates a DNS record inside the cluster called `redis` that points to our `redis` deployment. This DNS record is accessible to other deployments in the cluster.

4. Let's access our Redis pods to test some basic commands to store a value in our database. To do this, run the following command:

```
$ kubectl exec -it $(kubectl get pods -l run=redis
--output=jsonpath={..metadata.name}) -- redis-cli
```

The prompt will look like this: `127.0.0.1:6379>`.

5. Now, authenticate to the Redis database using the AUTH command, and then use `set` and `get` to create the a key with the value `1`. Finally, exit using the `exit` command. This simple test will look like this:

```
127.0.0.1:6379> AUTH YOUR_PASSWORD
OK
127.0.0.1:6379> set a 1
OK
127.0.0.1:6379> get a
"1"
127.0.0.1:6379> exit
```

With this, you stored the a key with the value 1. Now you have used Redis to store simple values. After running exit, you will exit to the Redis pod.

Now you have learned how to deploy a simple Redis deployment, it's time to deploy MongoDB in the next section.

Using a MongoDB document-oriented NoSQL database

MongoDB is a document-oriented NoSQL database. It stores its data as JSON documents. It also implements sharding techniques to distribute data across its nodes and uses the MapReduce technique for data aggregation. It's easy to use and uses low resources for single node scenarios. For our MongoDB deployment, we are going to use a ConfigMap to store custom configurations. In this case, our MongoDB configuration is set to expose its port across the network, but for the moment we are not using Secrets to simplify the deployment. In the *Using a PostgreSQL object-relational and SQL database* section, we are going to explore the use of secrets, but before that, let's follow the next steps to deploy MongoDB:

1. Deploy your custom configuration to enable clients to connect to MongoDB:

```
$ cat <<EOF | kubectl apply -f -
apiVersion: v1
kind: ConfigMap
metadata:
  name: mongo-configmap
data:
  mongod-conf: |
    dbpath=/var/lib/mongodb
    logpath=/var/log/mongodb/mongodb.log
    logappend=true
    bind_ip = 0.0.0.0
    port = 27017
    journal=true
    auth = true
EOF
```

This exposes MongoDB to listen on port 27017 across the network.

2. Create the deployment using the ConfigMap called `mongo-configmap`, the PersistentVolumeClaim, and the MONGO_INITDB_ROOT_USERNAME, MONGO_INITDB_ROOT_PASSWORD, and MONGO_INITDB_DATABASE variables, which set the initial root, user, and the password to connect to MongoDB as root or with your defined user:

```
$ cat <<EOF | kubectl apply -f -
apiVersion: apps/v1
kind: Deployment
metadata:
  labels:
    app: mongo
  name: mongo
spec:
  replicas: 1
  selector:
    matchLabels:
      app: mongo
  template:
    metadata:
      labels:
        app: mongo
    spec:
      containers:
      - image: arm64v8/mongo:4.4
        name: mongo
        env:
        - name: MONGO_INITDB_ROOT_USERNAME
          value: "admin"
        - name: MONGO_INITDB_ROOT_PASSWORD
          value: "YOUR_PASSWORD"
        - name: MONGO_INITDB_DATABASE
          value: "mydatabase"
        ports:
        - containerPort: 27017
        resources:
          limits:
            cpu: "0.5"
```

```
          memory: "200Mi"
      volumeMounts:
      - mountPath: "/data/db"
        name: mongo-storage
      - mountPath: /mongoconf
        name: config
    volumes:
      - name: config
        configMap:
          name: mongo-configmap
          items:
          - key: mongod-conf
            path: mongod.conf
      - name: mongo-storage
        persistentVolumeClaim:
          claimName: db-pv-claim
EOF
```

> **Important Note**
>
> Be aware that if you want to use a version of MongoDB greater than 5.0, you need a device with ARMv8.2-A or higher in order to use it. That's the reason to use MongoDB 4.4 for this example. MongoDB 4.4 is supported to run on ARMv8 processors such as a Raspberry Pi.

3. Now create the service that exposes your MongoDB deployment as a service accessible inside the cluster (MongoDB uses port 27017 to connect):

```
$ cat <<EOF | kubectl apply -f -
apiVersion: v1
kind: Service
metadata:
  labels:
    app: mongo
  name: mongo
spec:
  ports:
  - port: 27017
    protocol: TCP
```

```
        targetPort: 27017
     selector:
       app: mongo
     type: ClusterIP
  EOF
```

4. Access the pod that contains MongoDB to test whether you are able to write some data:

```
$ kubectl exec -it $(kubectl get pods -l app=mongo
--output=jsonpath={..metadata.name}) -- mongo -uadmin
-pYOUR_PASSWORD
```

Once you are inside the pod, change into mydatabase and insert the {"a":1} document in the mycollection collection using db.mycolletion.insert. Then, list the inserted document using db.mycollection.find. Finally, execute exit to finish the Mongo session. The commands and output of this execution will look like this:

```
> use mydatabase
switched to db mydatabase
> db.mycollection.insert({"a":1})
WriteResult({ "nInserted" : 1 })
> db.mycollection.find()
{ "_id" : ObjectId("622c498199789d3b03b20c45"), "a" : 1 }
> exit
Bye
```

These are some basic commands to use MongoDB, to have a quick start with Mongo.

Now you know how to deploy a simple MongoDB database in K3s, let's move on to learn how to use Postgres in the next section.

Using a PostgreSQL object-relational and SQL database

PostgreSQL is an object-relational database, used because of its strong reputation for reliability, feature robustness, and performance. It uses SQL to query its data. It's also commonly used for storing files or to store data used to create machine learning models. So, let's learn how to deploy PostgreSQL in a very simple way. To do this, follow the next steps:

1. For this example, let's use Kubernetes Secrets, and let's create the password as YOUR_PASSWORD to give an example of how to hide sensible information as passwords. For this, let's generate a Base64 encoding for your password with the following command:

```
$ echo "YOUR_PASSWORD"| tr -d "\n"  | base64
```

The output will look like this:

```
WU9VU19QQVNTV09SRA==
```

2. Use the previous output to create your Secret object using a YAML file. You can create the
    db-password Secret with this value using the following command:

```
$ cat <<EOF | kubectl apply -f -
apiVersion: v1
kind: Secret
metadata:
  name: db-password
data:
  password: WU9VU19QQVNTV09SRA==
EOF
```

3. Now create the Postgres deployment with the following command:

```
$ cat <<EOF | kubectl apply -f -
apiVersion: apps/v1
kind: Deployment
metadata:
  labels:
    app: postgres
  name: postgres
spec:
  replicas: 1
  selector:
    matchLabels:
      app: postgres
  template:
    metadata:
      labels:
        app: postgres
    spec:
      containers:
      - image: arm64v8/postgres:14.2
        name: postgres
        env:
```

```
        - name: PGDATA
          value: "/var/lib/postgresql/data/pgdata"
        - name: POSTGRES_PASSWORD
          valueFrom:
            secretKeyRef:
              name: db-password
              key: password
      ports:
      - containerPort: 5432
      resources:
        limits:
          cpu: "0.5"
          memory: "200Mi"
      volumeMounts:
      - mountPath: "/var/lib/postgresql/data"
        name: postgres-storage
    volumes:
    - name: postgres-storage
      persistentVolumeClaim:
        claimName: db-pv-claim
EOF
```

4. Now create the `postgres` service by running the following command:

```
$ cat <<EOF | kubectl apply -f -
apiVersion: v1
kind: Service
metadata:
  labels:
    app: postgres
  name: postgres
spec:
  ports:
  - port: 5432
    protocol: TCP
    targetPort: 5432
  selector:
```

```
      app: postgres
    type: ClusterIP
EOF
```

5. Access the pod that contains Postgres to test whether you can write some data. To do this, run the following command:

```
$ kubectl exec -it $(kubectl get pods -l app=postgres
--output=jsonpath={..metadata.name}) -- bash -c
"PGPASSWORD='YOUR_PASSWORD' psql -h postgres -U postgres"
```

6. The prompt will look like postgres=#. Next, you will find some example commands and their output. This commands will be used to test whether our database works.

    First, create the VALUE_TABLE table with the ID and VALUE fields:

```
postgres=# CREATE TABLE VALUE_TABLE (ID INT PRIMARY KEY
NOT NULL,VALUE INT NOT NULL);
CREATE TABLE
```

    Then insert a record with ID=1 and VALUE=123:

```
postgres=# INSERT INTO VALUE_TABLE (ID,VALUE) VALUES
(1,123);
INSERT 0 1
```

    Show the values:

```
postgres=# SELECT * FROM VALUE_TABLE;
 id | value
----+-------
  1 |   123
(1 row)
```

    Delete the table:

```
postgres=# DROP TABLE VALUE_TABLE;
DROP TABLE
```

    Exit from Postgres:

```
postgres=# exit
```

Now you've learned how to install and run basic commands with Postgres to store your data in this database, let's move on to learn about Neo4j, a graph NoSQL database, in the next section.

Using a Neo4j graph NoSQL database

Neo4j is a graph database that can be used to store relationships between objects. Neo4j uses **Cypher Query Language** (**CQL**), which is the equivalent of SQL for relational databases. Neo4j also represents data using nodes, relationships, properties, and labels in a visual way. It supports ACID operations and native graph storage and processing. It has great scalability and enterprise support. Because of the way it stores data, it can be used for IoT applications to query relationships between data. So now, let's install Neo4j by following the next steps:

1. Create the deployment for Neo4j:

```
$ cat <<EOF | kubectl apply -f -
apiVersion: apps/v1
kind: Deployment
metadata:
  labels:
    app: neo4j
  name: neo4j
spec:
  replicas: 1
  selector:
    matchLabels:
      app: neo4j
  template:
    metadata:
      labels:
        app: neo4j
    spec:
      containers:
      - image: arm64v8/neo4j
        name: neo4j
        env:
        - name: NEO4J_AUTH
          value: none
        ports:
            - containerPort: 7474
              name: http
            - containerPort: 7687
```

```
            name: bolt
          - containerPort: 7473
            name: https
        volumeMounts:
          - name: neo4j-data
            mountPath: "/var/lib/neo4j/data"
      volumes:
        - name: neo4j-data
          persistentVolumeClaim:
            claimName: db-pv-claim
EOF
```

In this deployment, we are using the NEO4J_AUTH variable with its value set to none, to use the non-authentication method, just to simplify this example. You can also explore how to use secrets and other options by modifying this configuration.

> **Important Note**
> If you delete the NEO4J_AUTH variable, Neo4j by default sets the user name and password to neo4j. Then, after logging in, a dialog box will ask you to change this password.

2. Create the service to expose the bolt, http, and https ports that Neo4j uses:

```
$ cat <<EOF | kubectl apply -f -
apiVersion: v1
kind: Service
metadata:
  labels:
    app: neo4j
  name: neo4j
spec:
  ports:
  - name: https
    port: 7473
    protocol: TCP
    targetPort: 7473
  - name: http
    port: 7474
    protocol: TCP
```

```
      targetPort: 7474
    - name: bolt
      port: 7687
      protocol: TCP
      targetPort: 7687
  selector:
      app: neo4j
  type: ClusterIP
EOF
```

3. Expose the `http` and `bolt` ports, before connecting to Neo4j with the browser. To do this, run the following commands in different terminals:

    ```
    $ kubectl port-forward service/neo4j 7474:7474
    $ kubectl port-forward service/neo4j 7687:7687
    ```

4. Open your browser at the page `http://localhost:7474`, choose **Authentication type**: **No authentication**, then click on the **Connect** button:

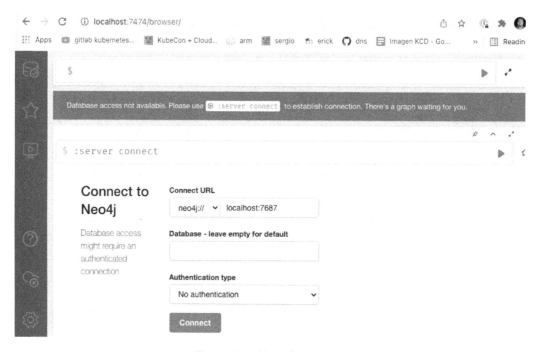

Figure 10.2 – Neo4j login page

Then you will see the Neo4j UI:

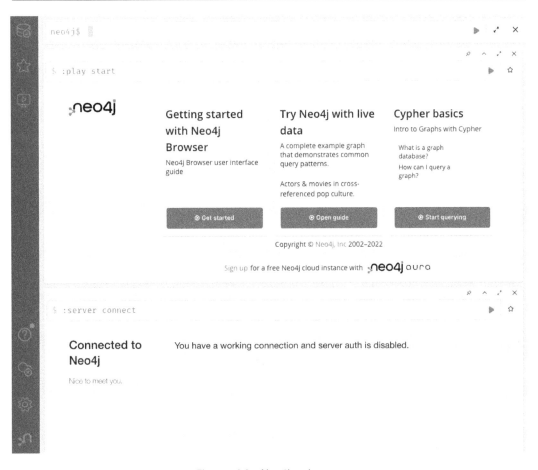

Figure 10.3 – Neo4j main page

5. Run a simple example in the Neo4j terminal located at the top of the browser as neo4j$. To do this, add the next commands and run them by clicking on the blue triangle button:

```
CREATE (IronMan:Hero{name: "Tony Stark"})
CREATE (Thanos:Villainous {name: "Thanos"})
CREATE (Thanos)-[r:ENEMY_OF]->(IronMan)
RETURN IronMan, Thanos
```

You will see that Neo4j visualizes the relationship between the Marvel characters:

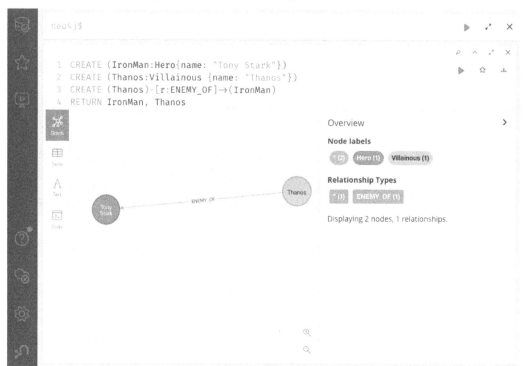

```
neo4j$                                                          ▶  ↗  ✕

                                                           ⌕  ∧  ↗  ✕
1  CREATE (IronMan:Hero{name: "Tony Stark"})               ▶  ☆  ↧
2  CREATE (Thanos:Villainous {name: "Thanos"})
3  CREATE (Thanos)-[r:ENEMY_OF]→(IronMan)
4  RETURN IronMan, Thanos
```

Overview   >

Node labels

`* (2)` `Hero (1)` `Villainous (1)`

Relationship Types

`* (1)` `ENEMY_OF (1)`

Displaying 2 nodes, 1 relationships.

Figure 10.4 – Neo4j graph visualization

Now you have learned how to use Neo4j with this basic example, let's move on to the summary of this chapter, about what we have learned.

Summary

In this chapter, we learned how to use CAP theorem to choose the right database to store data. This theorem helped us to take into consideration important guarantees when designing distributed data storage in a distributed system at the edge. In this chapter, we also learned about different relational and non-relational databases. We gained practical knowledge on how to set up and deploy various database paradigms such as relational, key-value, document-oriented, and graph databases. In the next chapter, we are going to focus on the time series database Prometheus, which stores data in the form of values and time and can be used to implement useful monitoring dashboards for devices at the edge.

Questions

Here are a few questions to validate your new knowledge:

- How can CAP theorem help you to decide which database to use according to your use case?

- How can you deploy MySQL in K3s?

- How can you deploy Redis in K3s?

- How can you deploy MongoDB in K3s?

- How can you deploy PostgreSQL on K3s?

- How can you deploy Neo4j on K3s?

- How can you use PersistentVolumeClaims to deploy a database on K3s?

- How can you use ConfigMaps and Secrets to deploy a database on K3s?

Further reading

You can refer to the following references for more information on the topics covered in this chapter:

- *Databases and Quick Overview of SQLite*: `https://medium.com/aiadventures/databases-and-quick-overview-of-sqlite-5b7d4f8f6174`

- *CAP Theorem for Databases: Consistency, Availability & Partition Tolerance*: `https://www.bmc.com/blogs/cap-theorem`

- Non-relational data and NoSQL: `https://aloa.co/blog/relational-vs-non-relational-database-pros-cons`

- CAP theorem: `https://devopedia.org/cap-theorem`

- *System design fundamentals: What is the CAP theorem?*: `https://www.educative.io/blog/what-is-cap-theorem`

- *What are the ACID properties of transactions and why do they matter in data engineering?*: `https://www.keboola.com/blog/acid-transactions`

- *SQL vs NoSQL Databases: What's The Difference?*: `https://www.bmc.com/blogs/sql-vs-nosql`

- *Traditional RDBMS to NoSQL database: New era of databases for big data*: `https://www.researchgate.net/publication/324922396_TRADITIONAL_RDBMS_TO_NOSQL_DATABASE_NEW_ERA_OF_DATABASES_FOR_BIG_DATA`

- MySQL client K8s: `https://gist.github.com/vishnuhd/b8686197f855c00fa734bc5f1fedf078`

- *Run a Single-Instance Stateful Application*: https://kubernetes.io/docs/tasks/run-application/run-single-instance-stateful-application

- *MySQL 8 Administrator's Guide*: https://www.packtpub.com/product/mysql-8-administrator-s-guide/9781788395199

- *Configuring Redis using a ConfigMap*: https://kubernetes.io/docs/tutorials/configuration/configure-redis-using-configmap

- *Redis Essentials*: https://www.packtpub.com/product/redis-essentials/9781784392451

- Kubernetes secrets: https://kubernetes.io/fr/docs/concepts/configuration/secret

- *Seven NoSQL Databases in a Week*: https://www.packtpub.com/product/seven-nosql-databases-in-a-week/9781787288867

- *How to use Kubernetes to deploy Postgres*: https://www.sumologic.com/blog/kubernetes-deploy-postgres

- *PostgreSQL 14 Administration Cookbook*: https://www.packtpub.com/product/postgresql-14-administration-cookbook/9781803248974

- *Internet of Things and Data: A Powerful Connection*: https://neo4j.com/news/internet-things-data-powerful-connection

- Why not SQLite: https://stackoverflow.com/questions/66950385/how-to-use-sqlite3-database-with-django-on-kuberenets-pod

- *Creating a Graph Application with Python, Neo4j, Gephi, and Linkurious.js*: https://linkurious.com/blog/creating-a-graph-application-with-python-neo4j-gephi-and-linkurious-js

Part 3:
Edge Computing Use Cases in Practice

In this part, you will learn how to use k3s and k3OS for different use cases, exploring complementary software and best practices for building an edge computing system.

This part of the book comprises the following chapters:

- *Chapter 11, Monitoring the Edge with Prometheus and Grafana*

- *Chapter 12, Communicating with Edge Devices across Long Distances Using LoRa*

- *Chapter 13, Geolocalization Applications Using GPS, NoSQL, and K3s Clusters*

- *Chapter 14, Computer Vision with Python and K3s Clusters*

- *Chapter 15, Designing Your Own Edge Computing System*

11
Monitoring the Edge with Prometheus and Grafana

One use case for edge computing is to monitor devices that get data about temperature, humidity, speed, noise, and so on. For this kind of use cases, monitoring would be critical. This chapter shows a simple use case of how to visualize data that comes from edge devices with sensors. This chapter presents a whole example of how to distribute and process data across the different layers of an edge computing system. This use case takes Prometheus and Grafana as the main components to visualize and store data from sensors and uses Mosquitto (an MQTT message broker) together with Redis to implement high availability queues to process data at the edge.

In this chapter, we're going to cover the following main topics:

- Monitoring edge environments
- Deploying Redis to persist Mosquitto sensor data
- Installing Mosquitto to process sensor data
- Processing Mosquitto topics
- Installing Prometheus, a time series database
- Deploying a custom exporter for Prometheus
- Configuring a DHT11 sensor to send humidity and temperature weather data
- Installing Grafana to create dashboards

Technical requirements

To deploy our databases in this chapter, you need the following:

- A single or multi-node K3s cluster that uses ARM devices with MetalLB and Longhorn storage installed. If you are using Raspberry Pi devices, you will need at least 4 GB of RAM and at least the 4B model. Each node has to have an Ubuntu ARM64 operating system in order to support the ARMv8 processor. This processor type is necessary for some deployments to run, because they use ARM64 container images.

- A Kubernetes cluster hosted in your public cloud provider (AWS, Azure, or GCP) or in your private cloud.

- A Raspberry Pi 4B with 2 or 4 GB for your edge device.

- A Keyes DHT11 sensor or similar connected to your edge device to read temperature and humidity.

- `kubectl` configured to be used in your local machine for your Kubernetes cloud cluster and your edge cluster, to avoid using the `--kubeconfig` parameter.

- A clone of the `https://github.com/PacktPublishing/Edge-Computing-Systems-with-Kubernetes/tree/main/ch11` repository, if you want to run the YAML configuration by using `kubectl apply` instead of copying the code from the book. Take a look at the `code` directory for Python source code and the `yaml` directory for YAML configurations located inside the `ch11` directory.

With this, you can deploy Prometheus and Grafana to start monitoring sensors data in edge environments.

Monitoring edge environments

Before starting to build our monitoring system, let's describe the system across the different layers of edge computing. For this, let's take a look at the following diagram:

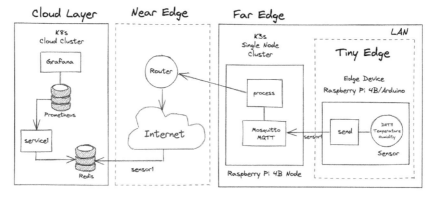

Figure 11.1 – Monitoring with edge devices

This diagram is divided into different layers. Let's describe the different components of this use case we want to implement:

- **Tiny edge**: Here you can find an edge device, in this case, a Raspberry Pi 4B. This Raspberry Pi works as an edge device that captures temperature and humidity data using a DHT11 sensor. Data is sent by running a small Python program called `send.py`. This file prepares the sensor to read data and sends the information to a queue in the Mosquitto broker.

- **Far edge**: Here is installed a K3s cluster using a Raspberry Pi 4B. Inside this cluster is installed Mosquitto. Mosquitto is a broker that uses the MQTT protocol and is designed to be lightweight, using few resources for processing. That's the reason Mosquitto is often used in edge and IoT scenarios. You can also find a process service that listens to a Mosquitto queue called `sensor1`. Every time the process detects new data, this data is sent to a Redis queue called `sensor1` in the cloud layer. The idea is that the deployment called `process` processes the information in the format to be shown in the cloud layer. With this, you are processing data near the edge; that is the goal of edge computing.

- **Near edge**: This is the home router that connects the edge device with the K3s cluster to process data. It is also the gateway to send data to the public Redis cluster in the cloud layer.

- **Cloud layer**: Here you can find a Kubernetes cluster with Prometheus, Grafana, and Redis installed. Prometheus is used as a time series database to store data from the edge sensors, and Grafana is used to visualize data. Every time data is generated in the far edge, it is sent to Redis. Redis is used to store data coming from tiny edge sensors in a temporary queue. In this way, Redis acts as backup storage if the communication fails in the far edge or if Prometheus is down. Technically speaking, `service1` is in charge of reading this data from the `sensor1` Redis queue and exporting it to Prometheus. Prometheus calls the `service1` service endpoint to get data. So, every time that Prometheus calls the `app1` endpoint, `service1` returns data stored in Redis in a format that Prometheus can consume. Finally, when data is stored in Prometheus, the data is visualized in real time in a Grafana dashboard.

As you can see, this small use case includes a whole interaction across the different edge computing layers. This use case pretends to be base code extensible to your own system needs. Now, let's start implementing our use case, beginning with deploying Redis to persist Mosquitto sensor data.

Deploying Redis to persist Mosquitto sensor data

To install our Redis to persist Mosquitto weather data, we are going to use Redis with persistence and a single list of messages. To deploy this Redis setup in your cluster, follow these steps:

1. Create the PersistentVolume to persist Redis data using the `/mnt/data` directory in the node:

    ```
    $ cat <<EOF | kubectl apply -f -
    apiVersion: v1
    ```

```
kind: PersistentVolume
metadata:
  name: db-pv-volume
  labels:
    type: local
spec:
  storageClassName: manual
  capacity:
    storage: 5Gi
  accessModes:
    - ReadWriteOnce
  hostPath:
    path: "/mnt/data"
EOF
```

2. Create a PersistentVolumeClaim using 5 GB of storage or more, depending on how many sensors and how much data you are processing:

```
$ cat <<EOF | kubectl apply -f -
apiVersion: v1
kind: PersistentVolumeClaim
metadata:
  name: db-pv-claim
spec:
  storageClassName: manual
  accessModes:
    - ReadWriteOnce
  resources:
    requests:
      storage: 5Gi
EOF
```

> **Important Note**
>
> You can use the longhorn class if Longhorn is installed in your system. For more information, see *Chapter 5, K3s Homelab for Edge Computing Experiments.*

3. Now let's create a ConfigMap to use a custom configuration with the password YOUR_PASSWORD and the /data directory to store Redis data:

```
$ cat <<EOF | kubectl apply -f -
apiVersion: v1
kind: ConfigMap
metadata:
  name: redis-configmap
  namespace: monitoring
data:
  redis-config: |
    dir /data
    requirepass YOUR_PASSWORD
EOF
```

4. Create the Redis deployment using the previous ConfigMap called redis-configmap. This ConfigMap is mounted as a volume and its content is available using the redis.conf file. It also uses a PersistentVolumeClaim called db-pv-claim and uses resource limits for CPU and memory. Let's create this deployment by running the following command:

```
$ cat <<EOF | kubectl apply -f -
apiVersion: apps/v1
kind: Deployment
metadata:
  labels:
    run: redis
  name: redis
  namespace: monitoring
spec:
  replicas: 1
  selector:
    matchLabels:
      run: redis
  template:
    metadata:
      labels:
        run: redis
    spec:
```

```
        containers:
        - name: redis
          image: redis:6.2
          command:
            - redis-server
            - /redisconf/redis.conf
          ports:
          - containerPort: 6379
          resources:
            limits:
              cpu: "0.2"
              memory: "128Mi"
          volumeMounts:
          - mountPath: "/data"
            name: redis-storage
          - mountPath: /redisconf
            name: config
        volumes:
          - name: config
            configMap:
              name: redis-configmap
              items:
              - key: redis-config
                path: redis.conf
          - name: redis-storage
            persistentVolumeClaim:
              claimName: db-pv-claim
    EOF
```

> **Important Note**
>
> You can use the `arm64v8/redis:6.2` image instead of `redis:6.2` if you plan to deploy Redis on an ARM node.

5. Now create the `redis` service, setting port `6379` in the configuration:

```
$ cat <<EOF | kubectl apply -f -
apiVersion: v1
```

```
kind: Service
metadata:
  labels:
    run: redis
  name: redis
  namespace: monitoring
spec:
  ports:
  - port: 6379
    protocol: TCP
    targetPort: 6379
  selector:
    run: redis
  type: ClusterIP
EOF
```

This service will be used by the exporter that Prometheus reads as `service1`.

6. Now create a `LoadBalancer` service called `redis-lb` to create a public load balancer that the `process` service can use to store data going from the far edge to the cloud layer:

```
$ cat <<EOF | kubectl apply -f -
apiVersion: v1
kind: Service
metadata:
  labels:
    run: redis
  name: redis-lb
  namespace: monitoring
spec:
  ports:
  - port: 6379
    protocol: TCP
    targetPort: 6379
  selector:
    run: redis
  type: LoadBalancer
EOF
```

This is going to create an external IP to access Redis.

7.   To get the public IP generated by the previous LoadBalancer service, run the following command:

```
$ EXTERNAL_IP="$(kubectl get svc redis-lb -n monitoring
-o=jsonpath='{.status.loadBalancer.ingress[0].ip}')"
```

This IP will be used by the deployment process.

Now our Redis is ready to be used in the far edge. Let's install Mosquitto to send sensor data to the sensor1 topic from Mosquitto.

Installing Mosquitto to process sensor data

Mosquitto is an open source broker that implements the MQTT protocol, and it's lightweight too. It was designed to be used with low-power sensors and devices. This makes Mosquitto suitable for edge computing and IoT applications. Mosquitto provides a lightweight channel of communication for edge devices and uses the publisher/subscriber pattern to send and read messages, but it is not persistent. We are going to use Redis later to give this missing temporary persistence for the data queues. Now, let's move to install Mosquitto in our edge cluster, located at the far edge. Remember that this single node cluster is using an ARM device. To deploy Mosquitto, follow these steps:

1.   Create a ConfigMap to listen over all the available network interfaces:

```
$ cat <<EOF | kubectl apply -f -
apiVersion: v1
kind: ConfigMap
metadata:
  name: mosquitto-configmap
data:
  mosquitto-config: |
    listener 1883 0.0.0.0
    allow_anonymous true
EOF
```

2.   Now create a deployment for Mosquitto, setting the ports to 1883 for the MQTT protocol and 9001 for HTTP requests. This deployment is going to use the previously created mosquitto-configmap:

```
$ cat <<EOF | kubectl apply -f -
apiVersion: apps/v1
kind: Deployment
```

```
metadata:
  labels:
    app: mosquitto
  name: mosquitto
spec:
  replicas: 1
  selector:
    matchLabels:
      app: mosquitto
  template:
    metadata:
      labels:
        app: mosquitto
    spec:
      containers:
      - name: mosquitto
        image: arm64v8/eclipse-mosquitto:2.0.14
        ports:
        - containerPort: 1883
          name: mqtt
        - containerPort: 9001
          name: http
        resources:
          limits:
            cpu: "0.2"
            memory: "128Mi"
        volumeMounts:
        - mountPath: /mosquitto/config
          name: config
      volumes:
      - name: config
        configMap:
          name: mosquitto-configmap
          items:
          - key: mosquitto-config
```

```
                  path: mosquitto.conf
    EOF
```

You can customize the amount of RAM and CPU that this deployment is using.

3. Now create a `ClusterIP` service to expose Mosquitto, so that other services inside the cluster can connect to Mosquitto to read messages:

```
$ cat <<EOF | kubectl apply -f -
apiVersion: v1
kind: Service
metadata:
  labels:
    app: mosquitto
  name: mosquitto
spec:
  ports:
  - name: mqtt
    port: 1883
    protocol: TCP
    targetPort: 1883
  - name: http
    port: 9001
    protocol: TCP
    targetPort: 9001
  selector:
    app: mosquitto
  type: ClusterIP
EOF
```

4. Now create a LoadBalancer service to expose Mosquitto, so that edge devices can connect to Mosquitto to publish messages with weather metrics. In this example, our device will publish in the `sensor1` topic:

```
$ cat <<EOF | kubectl apply -f -
apiVersion: v1
kind: Service
metadata:
  labels:
    app: mosquitto
```

```
        name: mosquitto-lb
      spec:
        ports:
        - name: mqtt
          port: 1883
          protocol: TCP
          targetPort: 1883
        - name: http
          port: 9001
          protocol: TCP
          targetPort: 9001
        selector:
          app: mosquitto
        type: LoadBalancer
      EOF
```

Now, let's deploy the `process` service that sends all the weather data stored in the Mosquitto topics to the Redis database located in the cloud layer.

Processing Mosquitto topics

We have to deploy the deployment called `process` using the `mqttsubs` container image, which sends the data published in Mosquitto to a public or private Redis instance in the cloud layer. Let's explore the code inside this container image:

```
import paho.mqtt.client as mqtt
import os
import redis
import sys


mqhost = os.environ['MOSQUITTO_HOST']
rhost = os.environ['REDIS_HOST']
rauth = os.environ['REDIS_AUTH']
stopic = os.environ['SENSOR_TOPIC']


def on_connect(client, userdata, flags, rc):
    client.subscribe(stopic)
```

```
def on_message(client, userdata, msg):
    r = redis.StrictRedis(host=rhost,\
        port=6379,db=0,password=rauth,\
        decode_responses=True)
    r.rpush(stopic,msg.payload)

client = mqtt.Client()
client.on_connect = on_connect
client.on_message = on_message
client.connect(mqhost, 1883, 60)
client.loop_forever()
```

> **Note**
>
> You can find the source of mqttsubs at https://github.com/sergioarmgpl/containers/tree/main/mqttsubs/src.

With this code, we get the necessary values to connect to Redis, the name of the topic that we are going to use. This value will be used to push sensor data into a Redis list. Finally, MOSQUITTO_HOST is where this service will be listened to. What this script basically does is it start listening to the SENSOR_TOPIC topic called sensor1 from Mosquitto, and when a message arrives, it is inserted into a Redis list with the same name in the cloud layer to persist the information temporarily. Redis uses port 6379 and is public but uses a password. Mosquitto is internally deployed on the far edge. This is how this service works.

To start deploying our process deployment, follow these steps:

1. Create a Secret to store the password to connect to Redis. Redis will be used as a way to store all the information coming from our Mosquitto deployment:

   ```
   $ cat <<EOF | kubectl apply -f -
   apiVersion: v1
   kind: Secret
   metadata:
     name: db-password
   data:
     password: WU9VU19QQVNTV09SRA==
   EOF
   ```

The value of the password corresponds to the output of the next command using base64 encoding:

```
$ echo "YOUR_PASSWORD" | tr -d '\n'  | base64
```

2. Create the `process` deployment that receives data coming from a Mosquitto topic and send it to the Redis service located in the cloud layer. For this run the following command:

```
$ cat <<EOF | kubectl apply -f -
apiVersion: apps/v1
kind: Deployment
metadata:
  labels:
    app: process
  name: process
spec:
  replicas: 1
  selector:
    matchLabels:
      app: process
  template:
    metadata:
      labels:
        app: process
    spec:
      containers:
      - image: sergioarmgpl/mqttsubs
        imagePullPolicy: Always
        name: mqttsubs
        env:
        - name: MOSQUITTO_HOST
          value: "mosquitto"
        - name: REDIS_HOST
          value: "192.168.0.242"
        - name: REDIS_AUTH
          valueFrom:
            secretKeyRef:
              name: db-password
              key: password
```

```
       - name: SENSOR_TOPIC
         value: "sensor1"
  EOF
```

The used variables are as follows:

- **MOSQUITTO_HOST:** This is the hostname where the Mosquitto deployment is listening.

- **REDIS_HOST:** This is the IP address assigned to the LoadBalancer service that exposes Redis in the cloud.

- **REDIS_AUTH:** This variable uses the db-password secret value to set the password to connect with Redis.

- **SENSOR_TOPIC:** This variable sets the Mosquitto topic to be listened to in order to get data from the sensors.

If you are using a private cloud, you might use an IP address like 192.168.0.242, for example. You can get this IP address by reading the *Deploying Redis to persist Mosquitto sensor data* section. Then, change the REDIS_HOST IP address to this value.

We have finished this section and have understood how data is processed. Let's continue deploying Prometheus service to store sensor data coming from the temporary Redis list.

Installing Prometheus, a time series database

Prometheus is a time series database that you can use to store your weather data. It's open source and it's suitable for edge devices. It can be deployed on ARM devices and it's very flexible to manage metrics and alerts. In this use case, we use Prometheus because of how flexible it is and the support it provides to store and visualize metrics. But we are going to use Grafana for visualizing data later. Now let's install Prometheus in our Kubernetes cloud cluster, following these steps:

1. Create the monitoring namespace, which will be used to install Prometheus and Grafana:

```
$ cat <<EOF | kubectl apply -f -
apiVersion: v1
kind: Namespace
metadata:
  name: monitoring
EOF
```

2. Create a ConfigMap that contains static configurations for Prometheus. In this case, we are going to create two services that insert data into Prometheus: one stores a counter and the weather data. The first service is called `service1` and the second `service2`. Each service uses port `5555`. Let's call this ConfigMap `prometheus-server-conf`. To create it, run the following command:

```
$ cat <<EOF | kubectl apply -f -
apiVersion: v1
kind: ConfigMap
metadata:
  name: prometheus-server-conf
  labels:
    name: prometheus-server-conf
  namespace: monitoring
data:
  prometheus.yml: |-
    global:
      scrape_interval: 5s
      evaluation_interval: 5s
      external_labels:
        monitor: 'codelab-monitor'
    scrape_configs:
      - job_name: 'MonitoringJob1'
        scrape_interval: 5s
        static_configs:
          - targets: ['service1:5555']
EOF
```

Targets are services that export data in the format that Prometheus can read. In this case, we are using two services. `service1` exports data from `sensor1`; this data is collected by Redis and transformed to be consumed by Prometheus. In this use case, we are going to use only `service1`, but you can create as many services as you want.

3. Now create the deployment for Prometheus, using the previous ConfigMap to configure Prometheus when its created, by running the following:

```
$ cat <<EOF | kubectl apply -f -
apiVersion: apps/v1
kind: Deployment
```

```yaml
metadata:
  name: prometheus-deployment
  namespace: monitoring
  labels:
    app: prometheus-server
spec:
  replicas: 1
  selector:
    matchLabels:
      app: prometheus-server
  template:
    metadata:
      labels:
        app: prometheus-server
    spec:
      containers:
        - name: prometheus
          image: prom/prometheus:v2.34.0
          args:
            - "--storage.tsdb.retention.time=12h"
            - "--config.file=/etc/prom/prometheus.yml"
            - "--storage.tsdb.path=/prometheus/"
          ports:
            - containerPort: 9090
          resources:
            requests:
              cpu: 500m
              memory: 500M
            limits:
              cpu: 1
              memory: 1Gi
          volumeMounts:
            - name: prometheus-config-volume
              mountPath: /etc/prom/
            - name: prometheus-storage-volume
              mountPath: /prometheus/
```

```
        volumes:
          - name: prometheus-config-volume
            configMap:
              defaultMode: 420
              name: prometheus-server-conf
          - name: prometheus-storage-volume
            emptyDir: {}
    EOF
```

This deployment listens on port 9090. This port is used to connect to Prometheus.

> **Important Note**
> You can use the same YAML to deploy Prometheus in a Kubernetes cluster deployed using a cloud provider, such as GCP, AWS, or Azure.

4. Now create a ClusterIP service that redirects port 9090 to port 8080 for Prometheus:

```
$ cat <<EOF | kubectl apply -f -
apiVersion: v1
kind: Service
metadata:
  creationTimestamp: null
  labels:
    app: prometheus-server
  name: prometheus-service
  namespace: monitoring
spec:
  ports:
  - port: 8080
    protocol: TCP
    targetPort: 9090
  selector:
    app: prometheus-server
  type: ClusterIP
EOF
```

5. Let's explore Prometheus by using `port-forward` to access the UI. For this, run the following command:

```
$ kubectl port-forward svc/prometheus-service 8080 -n
monitoring --address 0.0.0.0
```

6. Access `http://localhost:8080`; you will see the following page:

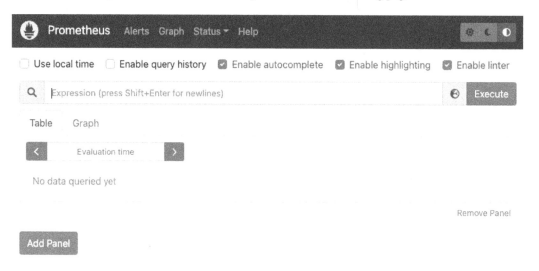

Figure 11.2 – Prometheus main page

7. Now go to the **Status** | **Targets** menu:

Figure 11.3 – Status menu

You will see the following page:

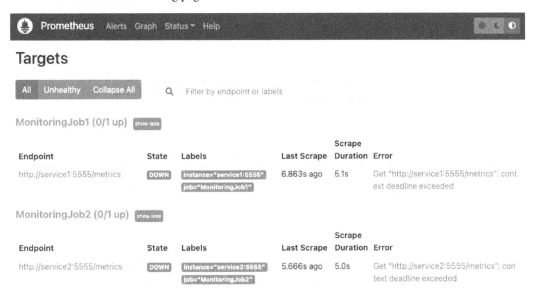

Figure 11.4 – Prometheus with targets down

On this page, you will see that the monitoring jobs are **down** at the moment. Because the services are not already created. After these monitoring services are created in the cluster, the state will change to **Up** using green color.

Now the Prometheus deployment is ready. Let's install our custom exporter in the cloud layer to export the temporary sensor data from our Redis list to Prometheus.

Deploying a custom exporter for Prometheus

After configuring all the components, you need to deploy the exporter that Prometheus calls to get data from Redis; this service will be called `service1`. Remember that Redis was being used to persist temporary data that comes from the Mosquitto topic on the far edge. Before deploying this service, let's understand the `exporter` container source code:

```
from flask import Response, Flask, request, jsonify
import prometheus_client
from prometheus_client import Gauge
import redis
import os
import sys
```

```python
import json

t = Gauge('weather_metric1', 'temperature')
h = Gauge('weather_metric2', 'humidity')

rhost = os.environ['REDIS_HOST']
rauth = os.environ['REDIS_AUTH']
stopic = os.environ['SENSOR_TOPIC']
r = redis.StrictRedis(host=rhost,\
        port=6379,db=0,password=rauth,\
        decode_responses=True)

@app.route("/metrics")
def metrics():
    data = r.lpop(stopic)
    values = json.loads(str(data).replace("\'","\""))
    t.set(int(values["temperature"]))
    h.set(int(values["humidity"]))
    res = []
    res.append(prometheus_client.generate_latest(t))
    res.append(prometheus_client.generate_latest(h))
    print({"processed":"done"},file=sys.stderr)
    return Response(res, mimetype="text/plain")

if __name__ == '__main__':
    app.run(host='0.0.0.0', port=5555, debug=True)
```

In this code made using Python, first we set the REDIS_HOST and REDIS_AUTH variables to connect to Redis and SENSOR_TOPIC to correspond to the list name in Redis where sensor data is stored. So, every time Prometheus calls the /metrics path, it extracts and returns one element inside the Redis list set with the value of SENSOR_TOPIC and returns a response in a format that Prometheus can read. For this, the code uses the prometheus_client library and sets two metrics using the Gauge metric type, which represents simple values. In this code, we are using two metrics: the first one is called weather_metric1, which contains the temperature values, and the second is weather_metric2, which contains humidity data. Once data is stored in Prometheus, it returns the JSON response {"processed":"done"}; after that, you can access this information in Prometheus. Alternatively, you can connect Prometheus to Grafana to create a new graph to show this data in real time.

> **Important Note**
>
> You can find the source of the exporter at `https://github.com/sergioarmgpl/` `containers/tree/main/exporter/src`.

Now let's deploy the exporter by following these steps:

1. Create the exporter by creating the `service1` deployment:

```
$ cat <<EOF | kubectl apply -f -
apiVersion: apps/v1
kind: Deployment
metadata:
  labels:
    app: service1
  name: service1
  namespace: monitoring
spec:
  replicas: 1
  selector:
    matchLabels:
      app: service1
  template:
    metadata:
      labels:
        app: service1
      annotations:
        prometheus.io/scrape: "true"
        prometheus.io/path: /metrics
        prometheus.io/port: "5555"
    spec:
      containers:
      - image: sergioarmgpl/exporter
        name: exporter
        env:
        - name: REDIS_HOST
          value: "redis"
        - name: REDIS_AUTH
```

```
              value: "YOUR_PASSWORD"
          - name: SENSOR_TOPIC
              value: "sensor1"
    EOF
```

You can use secrets instead of using the plain password in your YAML.

2. Now create the `service1` service:

```
$ cat <<EOF | kubectl apply -f -
apiVersion: v1
kind: Service
metadata:
  labels:
      app: service1
  name: service1
  namespace: monitoring
spec:
  ports:
  - port: 5555
      protocol: TCP
      targetPort: 5555
  selector:
      app: service1
  type: ClusterIP
EOF
```

Now if you return to your Prometheus targets, `service1` will appear as up and in green.

Now the exporter is running. It's time to configure the Python script in the edge device to get data coming from the DHT11 sensor and send it to the Mosquitto topic. Let's explore this in the next section.

Configuring a DHT11 sensor to send humidity and temperature weather data

Before you start using your edge device with a DHT11 sensor to send data, you need to follow these steps to configure it:

1. Install at least Ubuntu 20.04 LTS on your Raspberry Pi. You can check *Chapter 2, K3s Installation and Configuration*, and *Chapter 5, K3s Homelab for Edge Computing Experiments*, for more on this.

2. Configure your DHT11 sensor to send data to the Raspberry Pi. For this use case, we are going to use the DHT11 Keyes sensor, which comes from the Keystudio Raspberry Pi 4B Complete RFID Starter kit. This is a common sensor that you can find in other brands. This sensor gets the temperature and humidity. It often comes with three pins, which are *G = Ground*, *V = VCC*, and *S = Signal*. The way to connect is to connect G to a ground pin on the Raspberry and V to a 3V3 pin that powers the sensor with 3 volts. S, for signal, sends information to the Raspberry using a GPIO pin. In this case, you can use any free GPIO pin on the Raspberry; for this configuration, we are using the GPIO22 pin:

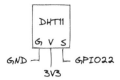

Figure 11.5 – DHT11 Keyes temperature and humidity sensor

3. Now, install the system and Python libraries that we need to run the sensor code in your edge device by running the following commands:

    • If `python3` is not installed in your Linux distribution, you can install it using the following command:

    ```
    $ sudo apt-get install python3 -y
    ```

    • Then continue installing the needed libraries:

    ```
    $ sudo apt-get install libgpiod2 git -y
    $ sudo python3 sensor.py
    $ sudo pip3 install adafruit-circuitpython-dht
    $ sudo pip3 install psutil
    $ sudo apt-get install i2c-tools
    ```

4. Clone the repository:

    ```
    $ git clone https://github.com/PacktPublishing/Edge-
    Computing-Systems-with-Kubernetes
    $ cd Edge-Computing-Systems-with-Kubernetes/ch11/code
    ```

5. Run the following:

    ```
    $ sudo python3 send.py
    ```

> **Important Note**
>
> Only run the send.py code inside your edge device until all the components of the use case are deployed.

Now you are starting to send data from your edge device. But what is happening inside the send. py code? Let's take a look:

```python
import time
import board
import adafruit_dht
import psutil
import paho.mqtt.client as mqtt
import sys

for proc in psutil.process_iter():
    if proc.name() == 'libgpiod_pulsein'
        or proc.name() == 'libgpiod_pulsei':
        proc.kill()

sensor = adafruit_dht.DHT11(board.D22)

mqhost="192.168.0.243"
client = mqtt.Client()
client.connect(mqhost, 1883, 60)
client.loop_start()

 def main():
    while True:
        t = sensor.temperature
        h = sensor.humidity
        client.publish("sensor1",\
        str({"t":int(t),"h":int(h)}))
        time.sleep(2)
try:
  main()
except KeyboardInterrupt:
  pass
```

```
finally:
  sensor.exit()
```

In this code, first, it's validated if the Raspberry Pi can read data from the GPIO pins. Then, by using the Adafruit library, we set the GPIO22 pin of the Raspberry Pi to read data from the sensor. After this, we set the Mosquitto host with the IP of the LoadBalancer service where Mosquitto is listening. Finally, we start a loop to read data with the `sensor` variable. This data is sent to the Mosquitto `sensor1` topic. The loop sends data every 2 seconds.

If you press *Ctrl + C*, the code stops and executes `sensor.exit()` to close the sensor and clean the state of the sensor. Finally, you are sending data. At this point, all the data passes across Mosquitto at the far edge and goes to Redis and Prometheus in the cloud layer. The only part that's missing is Grafana to visualize this data. For this, let's continue to the next section.

Installing Grafana to create dashboards

Grafana is a web application that you can use to visualize data from different data sources; it can also create alerts based on the data that you are visualizing. In our use case, Grafana will be used to visualize data that comes from Prometheus. Let's remember that Prometheus is listening to `service1`, to get data that comes from Mosquitto at the far edge. To deploy Grafana, follow these steps:

1. First, create a ConfigMap to configure your Grafana deployment:

    ```
    $ cat <<EOF | kubectl apply -f -
    apiVersion: v1
    kind: ConfigMap
    metadata:
      name: grafana-datasources
      namespace: monitoring
    data:
      prometheus.yaml: |-
        {
          "apiVersion": 1,
          "datasources": [
            {
              "access":"proxy",
              "editable": true,
              "name": "prometheus",
              "orgId": 1,
              "type": "prometheus",
    ```

```
              "url": "http://prometheus-service.monitoring.
    svc:8080",
              "version": 1
          }
        ]
      }
    EOF
```

This will be the default data source configured in your `grafana` deployment.

2. Let's create the `grafana` deployment by running the following:

```
$ cat <<EOF | kubectl apply -f -
apiVersion: apps/v1
kind: Deployment
metadata:
  name: grafana
  namespace: monitoring
spec:
  replicas: 1
  selector:
    matchLabels:
      app: grafana
  template:
    metadata:
      name: grafana
      labels:
        app: grafana
    spec:
      containers:
      - name: grafana
        image: grafana/grafana:8.4.4
        ports:
        - name: grafana
          containerPort: 3000
        resources:
          limits:
            memory: "1Gi"
```

```
                    cpu: "1000m"
                requests:
                    memory: 500M
                    cpu: "500m"
            volumeMounts:
                - mountPath: /var/lib/grafana
                  name: grafana-storage
                - mountPath: /etc/grafana/provisioning/
    datasources
                  name: grafana-datasources
                  readOnly: false
        volumes:
            - name: grafana-storage
              emptyDir: {}
            - name: grafana-datasources
              configMap:
                    defaultMode: 420
                    name: grafana-datasources
    EOF
```

3. Let's create the service:

```
$ cat <<EOF | kubectl apply -f -
apiVersion: v1
kind: Service
metadata:
  creationTimestamp: null
  labels:
    app: grafana
  name: grafana
  namespace: monitoring
spec:
  ports:
  - port: 3000
    protocol: TCP
    targetPort: 3000
  selector:
```

```
      app: grafana
   type: ClusterIP
EOF
```

4. Let's open the Grafana UI by running the following command:

    ```
    $ kubectl port-forward svc/grafana 3000 -n monitoring
    --address 0.0.0.0
    ```

5. Let's open the URL `http://localhost:3000`. When the login page appears, use the username `admin` and password `admin`, and click on the **Log in** button:

Figure 11.6 – Grafana login

6. After login, you will see the main page of Grafana:

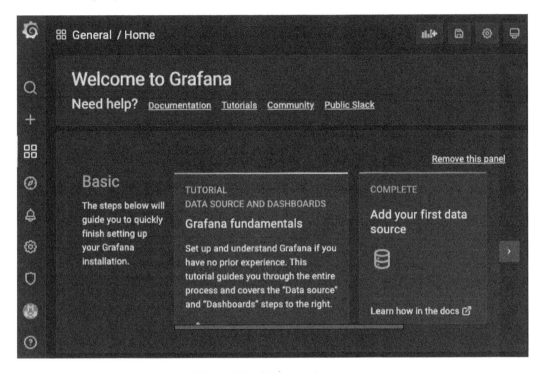

Figure 11.7 – Grafana main page

7. Click on **Configuration | Data sources**:

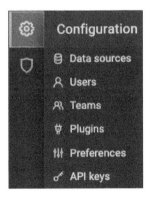

Figure 11.8 – Grafana configuration menu

8. Then, check whether the Prometheus data source exists:

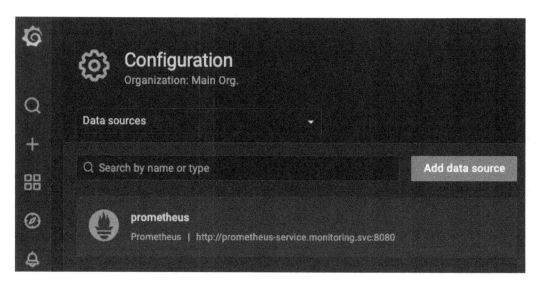

Figure 11.9 – Grafana data sources

Because of our ConfigMap configuration, our default data source will be `prometheus-service.monitoring.svc:8080`.

9. Now create a new folder or dashboard using the + icon. Let's create a folder first:

Figure 11.10 – Grafana Create menu

10. Now in the opened dialog fill the **Folder name** field with the value `Dashboard Sensors` to create a folder with this name, then click on the **Create** button:

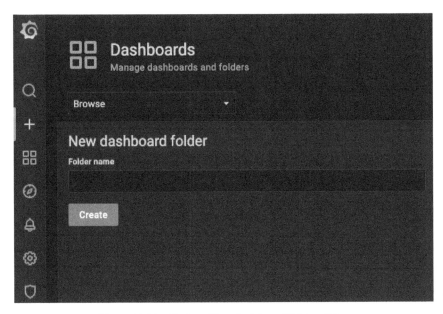

Figure 11.11 – Grafana New dashboard folder dialog

You can use this folder to save your dashboards and alerts if you want.

11. As in *Figure 11.10*, let's follow the same steps as for folders but this time click **Dashboard**. You will then see the page in *Figure 11.12*. Click on the **Add a new panel** button:

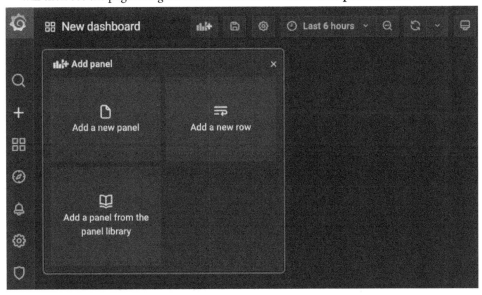

Figure 11.12 – Grafana Add panel page

12. In the next figure, you will see the settings to configure the new dashboard:

Figure 11.13 – Grafana New dashboard/Edit Panel page

Here you can configure this panel by setting the main part of the query. In this case, you have to write `weather_metric1` or `weather_metric2`. Here, `weather_metric1` gets the temperature and `weather_metric2` gets the humidity.

13. Set a time range to visualize data. Then, click on **Apply time range**:

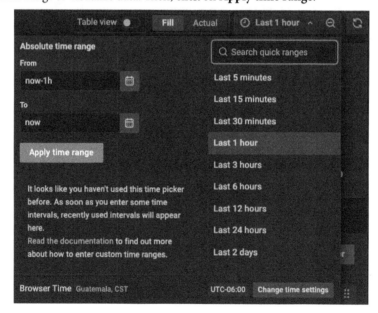

Figure 11.14 – Grafana Absolute time range dialog

14. Set the refresh time to 5 seconds in the next dialog, **Query options**:

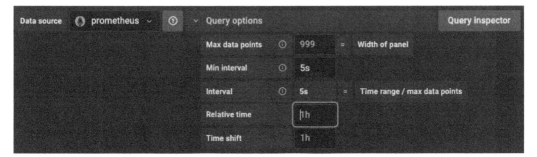

Figure 11.15 – Setting real-time data values

Alternatively, you can click on the **Refresh** icon:

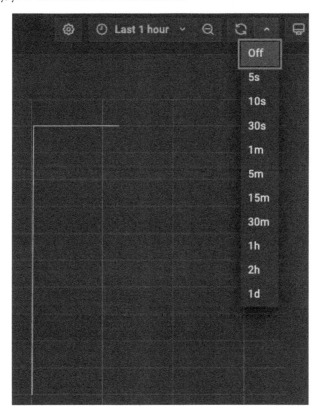

Figure 11.16 – Grafana setting refresh time

15. Then, click on the **Save** button and save the panel with any name and in a folder that you, for example with the previously folder created and using the name `Dashboard sensors` or `Temperature Sensor1`:

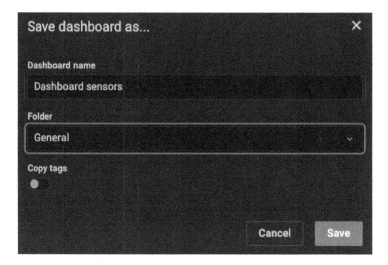

Figure 11.17 – Saving a new dashboard

16. You can also just apply the changes by clicking on the **Apply** button instead of the **Save** button:

Figure 11.18 – Applying changes to a new dashboard

17. Now you will see your dashboard:

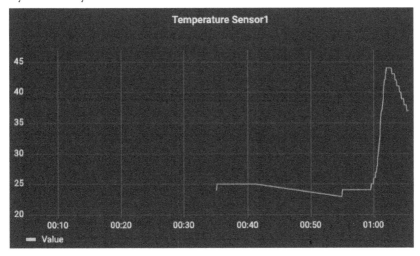

Figure 11.19 – Grafana Temperature Sensor1 dashboard

18. You can see the dashboards you have created by clicking on the **Search dashboards** icon:

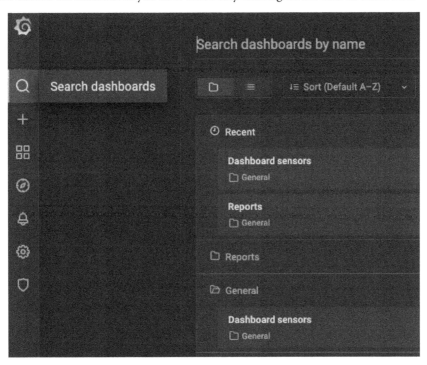

Figure 11.20 – Search dashboards

Now you can start visualizing the data that your edge device is generating, as shown in *Figure 11.18*. You can customize all the parameters to show the information according to your needs. You can also modify the code to add as many sensors as you want. We have now finished the chapter. Let's get a quick summary of what we learned.

Summary

In this chapter, we learned how monitoring can help us to visualize data at the edge, especially how to visualize data that comes from sensors, and how to build a basic use case scenario to extend for production use cases. To build this system, we used Prometheus as our time series database, Mosquitto as our basic way to store data from sensors, and Redis as a temporary queue to prevent the loss of our data from sensors. We also practiced how to build an edge computing system, using its different layers from the far edge to the cloud layer. This shows how important time series databases can be to manage sensor data and how tools such as Grafana can help to visualize it. This scenario can also be extended to farming, ocean and sea monitoring, animal populations, and so on. In the next chapter, we are going to continue with a similar scenario but applied to GPS and reading sensor data at long distances.

Questions

Here are a few questions to validate your new knowledge:

- How do I set up an edge device to capture sensor data?
- How do I use Prometheus to store data from sensors?
- How do I use Grafana to create custom graphs to visualize sensor data?
- How do I design a persistent system to manage sensor data using Mosquitto and Redis?
- How do I use Python to process and send sensor data?

Further reading

You can refer to the following references for more information on the topics covered in this chapter:

- Mosquitto official website: `https://mosquitto.org`
- Prometheus Python Client: `https://github.com/prometheus/client_python`
- How to set up Prometheus monitoring on a Kubernetes cluster `https://devopscube.com/setup-prometheus-monitoring-on-kubernetes`
- How to set up Grafana on Kubernetes: `https://devopscube.com/setup-grafana-kubernetes`

- Getting started with Prometheus: `https://prometheus.io/docs/prometheus/latest/getting_started`

- Using Prometheus and Grafana for IoT monitoring: `https://cloud.google.com/community/tutorials/cloud-iot-prometheus-monitoring`

- A step-by-step guide to setting up Prometheus Alertmanager with Slack, PagerDuty, and Gmail: `https://grafana.com/blog/2020/02/25/step-by-step-guide-to-setting-up-prometheus-alertmanager-with-slack-pagerduty-and-gmail`

12

Communicating with Edge Devices across Long Distances Using LoRa

Long Range (**LoRa**) is a wireless protocol that you can use to receive and send data over long distances using low-powered devices. You can use these edge devices with solar panels or other sources of energy. Sometimes, your edge devices use batteries and are not connected to a common power source like we often find in our houses. When you are crafting edge systems, you could use edge devices using sensors that you have to configure. You could use prototype hardware platforms such as Arduino or devices such as ESP32 microcontrollers or a Raspberry Pi. These devices support LoRa modules to bring communication capabilities to your device, which is crucial for sending and receiving data from devices. In this chapter, we are going to explore how to take advantage of the LoRa wireless protocol to send or receive data from long distances. We will continue expanding the options for monitoring edge devices as in the previous chapter but now using the LoRa wireless protocol.

In this chapter, we're going to cover the following main topics:

- LoRa wireless protocol and edge computing
- Deploying MySQL to store sensor data
- Deploying a service to store sensor data in a MySQL database
- Programming the ESP32 microcontroller to send sensor data
- Programming the ESP32 microcontroller to receive sensor data
- Visualizing data from ESP32 microcontrollers using MySQL and Grafana

Technical requirements

To deploy our databases in this chapter, you will need the following:

- A single or multi-node K3s cluster that can use ARM devices with MetalLB and Longhorn storage installed. This example will be tested using a Raspberry Pi 4B with 4 GB of RAM and using Ubuntu 20.04 or later for ARM 64-bit.

- A Kubernetes cluster hosted in your public cloud provider (AWS, Azure, GCP) or your private cloud.

- 2 ESP32 microcontrollers with the LoRa module installed. We are using the Heltec ESP32 + Lora v2 model; one to send and the other to receive data.

- Arduino IDE installed on your Mac. You can use Windows since it's pretty similar to configure, and it is also more stable when working with hardware.

- A USB 2.0 A-Male to Micro B cable for programming your ESP32 devices.

- A Keyes DHT11 sensor or similar connected to your edge device to read temperature and humidity.

- `kubectl` configured to be used in your local machine for your Kubernetes cloud cluster or your K3s cluster to avoid using the `--kubeconfig` parameter.

- Clone the GitHub repository at `https://github.com/PacktPublishing/Edge-Computing-Systems-with-Kubernetes/tree/main/ch12` if you want to run the YAML configurations using `kubectl apply` instead of copying the code from this book. Take a look at the `code` directory for Arduino source codes for Heltec devices and the `yaml` directory for YAML configurations. These are located inside the `ch12` directory.

Now, let's understand how our scenario of using LoRa devices, Prometheus, and Grafana is going to work.

LoRa wireless protocol and edge computing

LoRa refers to a radio modulation technique for long distances, and together with LoRaWAN, it defines a network protocol that can be used to interconnect devices. LoRaWAN is also a network architecture that uses a start-of-start topology in which the gateway relays messages between edge devices. LoRa uses three popular frequencies: 433, 868, and 915. 433 is sometimes used outdoors. 868 is used in Europe and 915 is used in America. LoRaWAN has gateway devices that can connect LoRa networks to the internet. LoRa is designed for low power, which is why LoRa is used for applications in IoT to interconnect devices across long distances.

As we know, the goal of edge computing is to process data near the source. Therefore, LoRa allows us to implement edge computing and interconnect devices for long distances without using a lot of energy. Some use cases include agriculture, buildings, supply chain, logistics, geo localization applications, and more. Some common devices that support LoRa are Heltec ESP32 devices, which are designed for low power consumption. We are going to focus on configuring a Heltec ESP32 device with LoRa support in this chapter.

To start our use case implementation of using LoRa on the tiny edge to interact with a far edge Kubernetes cluster, let's explore the following diagram:

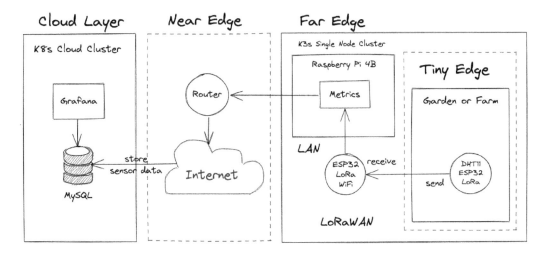

Figure 12.1 – Monitoring with ESP32 devices and LoRaWAN

This diagram is divided into different layers. You can see how the data flows between the far edge, where LoRa communication is implemented, to the cloud layer. But first, let's describe the different components of this use case that we want to implement:

- **Tiny edge**: In this layer, we are going to find the Heltec ESP32 devices, which you can classify as devices that send or receive data. This device sends data, reads data from the DHT11 sensor, and sends the information in JSON format using the LoRa protocol. The other devices read the information, send it across LoRa, and send it to the cluster on the far edge using a post request in a LAN. You can add as many devices to send data as you want.

- **Far edge**: Here, you can find a single or multi-node K3s cluster using ARM devices. This cluster provides the metrics service, which receives data in JSON format from the tiny edge. Once the data is received, the metrics service writes this data to the cloud layer in the MySQL deployment inside a Kubernetes cluster provisioned in the cloud provider. This could be Amazon, GCP, Azure, and so on.

- **Near edge**: This layer contains the local router that connects the local network to the internet. Keep in mind that the cluster in the far edge works as a gateway to send data from the LoRa network to the internet.

- **Cloud layer**: Here, you can find the Kubernetes cluster, which contains MySQL and Grafana. MySQL stores data coming from your local sensors, while Grafana uses MySQL to create dashboards using your sensor data.

In summary, all your sensor data is coming from ESP32 devices, some equipped with sensors. These devices send and receive data using the LoRa protocol. When a receiver device receives information, it's transformed into JSON format and then sent to a Kubernetes service located on the far edge. After this service receives this information, it's forwarded to the cloud layer and stored in a MySQL database. MySQL is used by Grafana to show sensor data collected at the edge in real time. Now, let's deploy our MySQL databases to store data.

Deploying MySQL to store sensor data

Before you can store data from your devices using LoRa, you must deploy your database. For this, we are going to use MySQL. MySQL is a pretty popular database that you can use to store data from your sensor. The main advantage of using MySQL is that it is well documented, and you can find a lot of examples on the internet. For our deployment, we are going to use a `PersistentVolumeClaim` and the `mysql:8.0.28-oracle` image. Even if you decide to deploy your MySQL over the cloud or locally at the edge, you must use a `LoadBalancer` service so that you have an endpoint for the service that is going to store all sensor data. Our MySQL database will be deployed in the default namespace to simplify the implementation. To deploy our MySQL database, follow these steps:

1. Create a `PersistentVolumeClaim` with 5 GB of storage:

    ```
    $ cat <<EOF | kubectl apply -f -
    apiVersion: v1
    kind: PersistentVolumeClaim
    metadata:
      name: db-pv-claim
    spec:
      accessModes:
        - ReadWriteOnce
      resources:
        requests:
          storage: 5Gi
    EOF
    ```

> **Important Note**
>
> Don't forget to use ConfigMaps and Secrets for a more secure and advanced configuration. You can explore *Chapter 10, SQL and NoSQL Databases at the Edge*, for more details.

2. Now, let's deploy our MySQL database. Our deployment is going to use the previous `PersistentVolumeClaim`, called `db-pv-claim`, for this run:

```
$ cat <<EOF | kubectl apply -f -
apiVersion: apps/v1
kind: Deployment
metadata:
  name: mysql
spec:
  selector:
    matchLabels:
      app: mysql
  strategy:
    type: Recreate
  template:
    metadata:
      labels:
        app: mysql
    spec:
      containers:
      - image: mysql:8.0.28-oracle
        name: mysql
        env:
        - name: MYSQL_DATABASE
          value: sensor_data
        - name: MYSQL_USER
          value: lora_mysql
        - name: MYSQL_PASSWORD
          value: lora123-
        - name: MYSQL_ROOT_PASSWORD
          value: lora123-
        ports:
```

```
        - containerPort: 3306
          name: mysql
        volumeMounts:
        - name: mysql-persistent-storage
          mountPath: /var/lib/mysql
      volumes:
      - name: mysql-persistent-storage
        persistentVolumeClaim:
          claimName: db-pv-claim
EOF
```

In this deployment, we are using some environment variables:

- MYSQL_DATABASE: Creates an initial database

- MYSQL_USER: Creates a super admin user for the database defined in MYSQL_DATABASE

- MYSQL_PASSWORD: Sets a password for the defined user in the MYSQL_USER variable

- MYSQL_ROOT_PASSWORD: Sets a password for the root user

> **Important Note**
>
> If you are using a multi-node cluster, use the nodeSelector option to prevent issues with the provisioned PersistentVolumeClaim.

3. Now, we need a ClusterIP service. This will be used inside Grafana to configure this MySQL database as a data source:

```
$ cat <<EOF | kubectl apply -f -
apiVersion: v1
kind: Service
metadata:
  creationTimestamp: null
  name: mysql
spec:
  ports:
  - port: 3306
    protocol: TCP
    targetPort: 3306
  selector:
    app: mysql
```

```
  type: ClusterIP
status:
  loadBalancer: {}
EOF
```

4. We also need a LoadBalancer service to expose MySQL. This service will be used to expose MySQL to the outside world. This could be over the internet or using an IP address inside your local network. Regardless, the provisioned load balancer IP address will be used inside your ESP32 devices. These ESP32 devices are going to send information to this endpoint, using our metrics service to finally store sensor data in MySQL. Let's create this LoadBalancer service:

```
$ cat <<EOF | kubectl apply -f -
apiVersion: v1
kind: Service
metadata:
  creationTimestamp: null
  name: mysql-lb
spec:
  ports:
  - port: 3306
    protocol: TCP
    targetPort: 3306
  selector:
    app: mysql
  type: LoadBalancer
status:
  loadBalancer: {}
EOF
```

Now that MySQL is running, we have to create a table to store sensor data. To do so, follow these steps:

1. Create a MySQL CLI client to run some commands to create the table where data will be stored:

```
$ kubectl run client -it --rm --image=mysql:8.0.28-oracle
-- bash
```

Once you are inside, run the following command. This will ask you for a password. Use lora123- as your password:

```
$ mysql -u lora_mysql -h mysql -p
```

The prompt will change to something similar to mysql>.

2. Create the metric table and include the `device`, `temperature_c`, `temperature_f`, `humidity`, and `time` fields:

```
use sensor_data;
CREATE TABLE metric (device INT NOT NULL,temperature_c
DECIMAL(4,2),temperature_f DECIMAL(4,2) NOT NULL,humidity
DECIMAL(4,2) NOT NULL, time DATETIME NOT NULL);
```

First, we must select the `sensor_data` database to create the table inside using the `use` command. Then, we must create the table metric using the `CREATE TABLE` command. We configure it so that each field has to have values. We use `DECIMAL(4,2)`, which means 4-2 = 2 integer numbers and 2 decimals. We store data using the format used by the `now()` MySQL function as `MONTH/DAY/YEAR HOUR:MINUTE:SECOND`.

Here is a small explanation of what each field contains:

- `device`: This represents the number of the ESP32 Lora device that sends sensor data. This could be a number greater than 0.

- `temperature_c`: This is the temperature measured in Celsius.

- `temperature_f`: This is the temperature measured in Fahrenheit.

- `humidity`: Ambient humidity, measured as a percentage.

3. Exit the MySQL client using the `quit` command inside MySQL.

4. Exit the client using the `exit` command. After exiting, the pod will be deleted.

5. Get the MySQL IP address provisioned in the `LoadBalancer` service. To do so, run the following commands:

```
$ MYSQL_IP="$(kubectl get svc mysql-lb -o=jsonpath='{.
status.loadBalancer.ingress[0].ip}')"
$ echo $MYSQL_IP
```

The `echo` command is going to show the IP address of your MySQL.

Now that our MySQL database has been deployed and is ready to be used, let's deploy our metrics application on the far edge to store data in this MySQL database.

Deploying a service to store sensor data in a MySQL database

For this scenario, we need to deploy a service to store data in the previously deployed MySQL. We are going to call this `metrics`. The `metrics` service contains the following code:

```
from flask import Flask, request
import mysql.connector
```

```python
import os

app = Flask(__name__)

@app.route('/')
def hello_world():
    return 'It works'

def insert(data):
    conn = mysql.connector.connect(
     host=os.environ['HOST'],
     user=os.environ['MYSQL_USER'],
     password=os.environ['MYSQL_PASSWORD'],
     database=os.environ['MYSQL_DATABASE']
    )
    cursor = conn.cursor()
    sql = "INSERT INTO metric "+\
          "(device,temperature_c,"+\
          "temperature_f,humidity,time) "+\
          "VALUES (%s,%s,%s,%s,now());"
    val = (data["d"],data["t"],data["t_f"],data["h"])
    cursor.execute(sql,val)
    conn.commit()
    cursor.close()
    conn.close()

@app.route('/device',methods = ['POST'])
def device():
    data = request.json
    print(data)
    #Process data in some way
    t_farenheit = float(data["t"])*(9/5)+32
    data["t_f"] = t_farenheit
    insert(data)
    return "processed"
```

```
if __name__ == '__main__':
    app.run(host='0.0.0.0', port=3000, debug=True)
```

This code has two endpoints:

- /: This is only a test URL.

- /device: This gets data from POST requests and writes it to MySQL by calling the `insert` function.

It also uses the `insert(data)` function to insert the data into the MySQL deployed in the cloud layer. This function takes the data coming from LoRaWAN and recalculates the temperature in Fahrenheit. Once stored, the data returns the process word.

This script also uses the following environment variables:

- HOST: Defines the IP address where MySQL is listening

- MYSQL_USER: The user to connect to the database

- MYSQL_PASSWORD: The password used to connect to the database

- MYSQL_DATABASE: The database name where `metrics` is going to store data

Our deployment in Kubernetes must have these variables set to work properly, without errors.

> **Important Note**
> You can check the code on how to build a container based on this example at `https://github.com/sergioarmgpl/containers/tree/main/metric`.

Now that we've looked at the code of the `metrics` service, let's deploy `metrics` to start storing sensor data in this database. For this, follow these steps:

1. Deploy our `metrics` application so that it's running:

```
$ cat <<EOF | kubectl apply -f -
apiVersion: apps/v1
kind: Deployment
metadata:
  creationTimestamp: null
  labels:
    app: metrics
  name: metrics
spec:
```

```
    replicas: 1
    selector:
      matchLabels:
        app: metrics
    strategy: {}
    template:
      metadata:
        creationTimestamp: null
        labels:
          app: metrics
      spec:
        containers:
        - image: sergioarmgpl/metric
          name: metric
          env:
          - name: HOST
            value: "192.168.0.240"
          - name: MYSQL_USER
            value: "lora_mysql"
          - name: MYSQL_PASSWORD
            value: "lora123-"
          - name: MYSQL_DATABASE
            value: "sensor_data"
          resources: {}
  status: {}
  EOF
```

In this deployment, we are using the following values for the environment variables:

- HOST: This is the IP address of the LoadBalancer service that was created for our MySQL – that is, 192.168.0.240. This will be the IP address that was returned in the last step of the previous section.

- MYSQL_USER: The MySQL user. In this case, this is lora_mysql.

- MYSQL_PASSWORD: The password for lora_mysql. In this case, this is lora123-.

- MYSQL_DATABASE: The name of the MySQL database where sensor data will be stored. In this case, this is sensor_data.

Important Note

You can customize all these values to fit your needs. Remember that you can use ConfigMaps or Secrets to secure your deployments. We are using hard-coded values just to simplify the implementation. Check out *Chapter 10, SQL and NoSQL Databases at the Edge*, for this kind of configuration.

---

2. Now, let's create our `LoadBalancer` service for the `metrics` deployment. The provisioned IP address will be hard-coded inside the code of our ESP32 devices. To create the service, run the following code:

```
$ cat <<EOF | kubectl apply -f -
apiVersion: v1
kind: Service
metadata:
  creationTimestamp: null
  name: metrics
spec:
  ports:
  - port: 3000
    protocol: TCP
    targetPort: 3000
  selector:
    app: metrics
  type: LoadBalancer
status:
  loadBalancer: {}
EOF
```

3. To obtain the provisioned IP address from the `metrics` service, run the following commands:

```
$ METRICS_IP="$(kubectl get svc metrics -o=jsonpath='{.
status.loadBalancer.ingress[0].ip}')"
$ echo METRICS_IP
```

The `echo` command will show the IP address of our `metrics` application. Take note of this value since it will be used to program our ESP32 devices. Let's assume that this value is `192.168.0.241` for this scenario.

Now that we have deployed our `metrics` service on the far edge, let's configure our ESP32 devices so that they can send and receive data.

Programming the ESP32 microcontroller to send sensor data

ESP32 is a low-cost, low-power microcontroller chip and the successor of the ESP8266 microcontroller. In this chapter, we will be using the Heltec ESP32 + LoRa, which is an ESP32 microcontroller plus the capability to use the LoRa wireless protocol. This microcontroller can also send and receive data using the LoRa wireless protocol using the integrated SX1276 chip in this Heltec dashboard.

Before configuring our device, we have to do the following:

1. Connect the DHT11 sensor to the Heltec ESP32 + LoRa device.

2. Install the USB to UART bridge driver.

3. Install and configure Arduino IDE to program the Heltec ESP32 + LoRa device.

4. Flash the Heltec ESP32 + Lora device.

So, let's get started by connecting a DHT11 sensor to our Heltec ESP32 + Lora device.

Configuring Heltec ESP32 + LoRa to read DHT11 sensor data

Heltec devices are often used for IoT, and ESP32 is a very popular device for IoT and LoRa implementations. You can find the official documentation for Heltec devices at `https://heltec-automation-docs.readthedocs.io/en/latest`. For our LoRa implementation, we are going to use the following diagram:

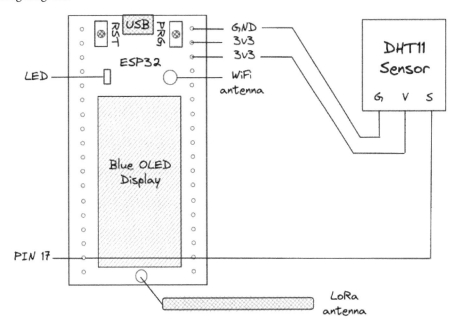

Figure 12.2 – Heltec ESP32 reading data from the DHT11 schema

To connect your Heltec ESP32 to your DHT11 sensor, follow these steps. This is the sender device:

1. Connect your LoRa antenna. This could affect the transmission range if the antenna is not connected.
2. Connect your power source to the USB connection. You must use a USB 2.0 A-Male to Micro B cable.
3. Connect your `PIN 17` with a wire to the `S` input in the DHT11 sensor.
4. Connect the GND (ground) to the `G` input in the DHT11 sensor.
5. Connect one of the 3V3 volt outputs to the `V` input in the DHT11 sensor.

For your receiver device, just follow these steps:

1. Connect your LoRa antenna.
2. Connect your power source to the USB connection as your sender device.

> **Important Note**
>
> Remember that you have to use a power source supply or a battery that gives you 3.5 or 5 volts. To learn more, check out `https://heltec-automation-docs.readthedocs.io/en/latest/esp32/index.html`.

With that, your sender and receiver devices are ready to upload some code. Now, let's install all the software that we need to upload some code into our devices.

Installing the USB to UART bridge driver

When installing the generic SiLabs CP210X driver that installs support for the USB to UART bridge, this driver is going to recognize your device on your computer. In this way, your Heltec device can connect to the computer and interact with Arduino IDE using serial communication.

To install this driver, you can visit the following link for detailed instructions: `https://heltec-automation-docs.readthedocs.io/en/latest/general/establish_serial_connection.html`. There, you can find the latest documentation to install the driver on Windows and Mac.

These instructions consist of downloading various drivers:

- For Windows: `https://www.silabs.com/documents/public/software/CP210x_Windows_Drivers.zip`

- For Mac: `https://www.silabs.com/documents/public/software/Mac_OSX_VCP_Driver.zip`

You must follow the wizard to install them, depending on your system. Once you have installed the driver, you can connect your device to your system and check if it was detected. For Mac, you can execute the following command:

```
$ ls /dev | grep cu | grep 'usbserial\|UART'
```

You will see an output similar to the following:

```
cu.SLAB_USBtoUART
cu.usbserial-0001
```

This means that your device was detected. Some common problems may occur where your device can't be recognized. This will be because of the cable that you are using; try to find a cable in an optimal condition for your computer to detect the device.

For Windows, you will see that your device appears in the Hardware manager in the part of ports. Then, the device will appear like so:

- Silicon Labs CP210X USB to UART bridge (COM3)

This means that your Heltec device was detected successfully.

Now that your system recognizes the device, it's time to install Arduino IDE to upload some code inside your Heltec device.

Installing Arduino IDE

Arduino IDE is a piece of software that you can use to upload code to your boards. In this case, we are using a board designed by Heltec, which is the one we called the Heltec device. To start using Arduino IDE, follow these steps:

1. Download Arduino IDE by going to `https://www.arduino.cc/en/software`. This will depend on which operating system are you using. You can choose between Windows, Linux, or Mac. In this chapter, we are going to cover just Mac. We will use Arduino 1.8.19.

> **Important Note**
>
> You can also follow the official page of Heltec, which explains how to install the Heltec driver and Arduino IDE for Windows and Mac. The quick start link is `https://heltec-automation-docs.readthedocs.io/en/latest/esp32/quick_start.html`.

2.   Open Arduino IDE by clicking on its icon on your desktop or inside Launchpad on Mac.

3.   Click **File | Preferences** and paste `https://github.com/Heltec-Aaron-Lee/WiFi_Kit_series/releases/download/0.0.5/package_heltec_esp32_index.json` inside the **Additional Boards Manager URLs** field. Then, click **OK**. The new ESP32 board will be loaded in Arduino IDE:

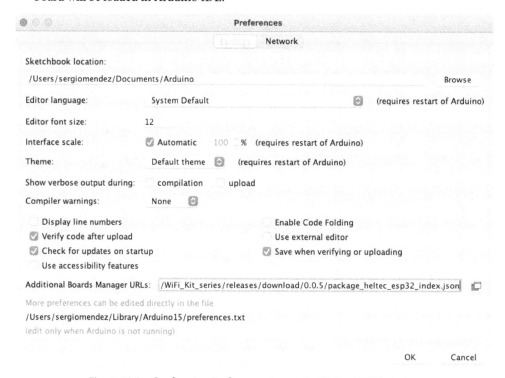

Figure 12.3 – Configuring Preferences to use the Heltec ESP32 device

4. Now, go to **Tools** | **Board** | **Boards Manager**:

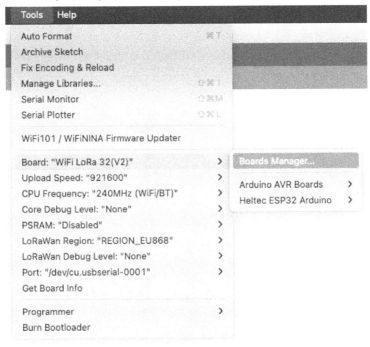

Figure 12.4 – The Boards Manager menu

5. Search for `heltec` in the new pop-up dialog, then click **Install** to install it:

Figure 12.5 – Searching for heltec on Boards Manager

You will see something similar to the following:

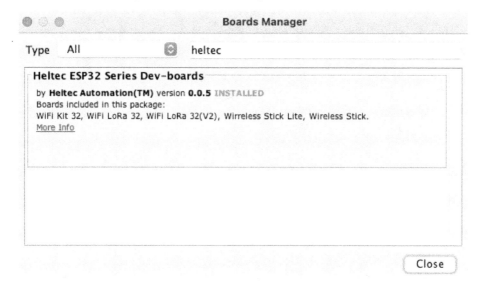

Figure 12.6 – Heltec board installed via Boards Manager

6. Now, select the board by going to **Tools** | **Board** | **Heltec ESP32 Arduino** and select **WiFi LoRa 32(V2)**:

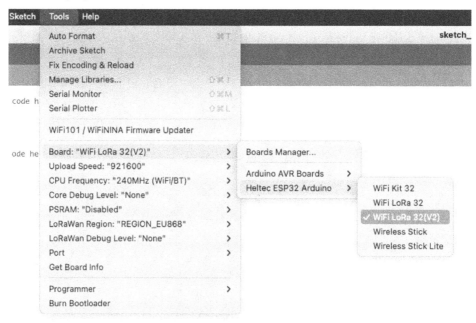

Figure 12.7 – Setting WiFi LoRa 32(V2) as the default board

7. Now, repeat this process by going to **Tools** | **Manage Libraries**, searching for the **DHT sensor** library from Adafruit, and choosing **Heltec ESP32 Dev-Boards**. Make sure you install it.

With that, our Arduino IDE is ready to be used. Now, let's learn about some configurations that you will need in case of errors.

Troubleshooting Arduino IDE when using Heltec ESP32 + LoRa

macOS has some challenges, depending on your Mac version, but you can fix them.

One is the esptool Python library. To fix it, follow these steps:

1. Copy your current `esptool.py` file inside the `tools` folder. The command will look as follows:

```
$ cp /Users/<YOUR_USER>/Library/Arduino15/packages/
Heltec-esp32/hardware/esp32/<X.X.X> /tools/esptool.py /
Users/ <YOUR_USER> /Library/Arduino15/packages/Heltec-
esp32/tools/esptool_py/<X.X.X>/
```

2. Change the permissions for the `esptool.py` file:

```
$ chmod +x esptool.py
```

Run the `esptool.py` file:

```
$ ./esptool.py
```

3. Sometimes, you have to install the serial library if the previous command returns an error. For this, you have several options. One is to install the library from scratch by going to `https://github.com/pyserial/pyserial/releases`. In this case, we are using version 3.4. For this run, the following commands:

```
$ wget https://github.com/pyserial/pyserial/archive/refs/
tags/v3.4.zip
$ sudo python setup.py install
```

4. Then, try again if the `./esptool.py` command returns errors.

5. Depending on which Python version is installed on your computer, you can try this other alternative:

```
$ sudo pip install pyserial or
$ sudo pip3 install pyserial
```

6. The last alternative is to use `easy_install` to install the pyserial library. To do so, run the following command:

```
$ sudo easy_install pyserial
```

> **Important Note**
>
> macOS Monterrey deletes Python 2.7 by default, so you have to install this Python version. You could stay with Python 3, but you have to open Arduino IDE with the `open /Applications/Arduino.app` command. You can find a more detailed way to fix these problems by watching the following video: `https://www.youtube.com/watch?v=zkyoghpT8_U`.

Another problem to fix, depending on your Arduino version, is that Heltec installs its Wi-Fi library. So, when you try to compile and upload the program, sometimes, you will see some errors. To avoid these errors, you have two options:

1. Uninstall the default Arduino Wi-Fi library by going to **Tools | Manage libraries**. Then, find the Wi-Fi library from Arduino and uninstall it.

2. Remove or rename the default Arduino Wi-Fi library folder by using the following commands:

```
$ cd /Applications/Arduino.app/Contents/Java/libraries
$ mv libraries/WiFi
```

> **Important Note**
>
> On Windows, the installation is smooth, so you won't have to fix this kind of issue.

Now, it is time to upload some code to your devices using Arduino IDE.

Uploading code to the ESP32 microcontroller to send sensor data

Now, let's upload our code to our Heltec devices. Let's start with the sender device. This device is going to capture data from the DHT11 sensor and send it to the receiver device using the LoRa wireless protocol. Let's create a new file by going to **File | New**. By default, you will see something similar to the following:

```
void setup() {
  // put your setup code here, to run once:
}

void loop() {
```

```
  // put your main code here, to run repeatedly:
}
```

Now, replace it with the following code:

```
#include "heltec.h"
#define BAND    915E6

#include "DHT.h"
#define DHTPIN 17
#define DHTTYPE DHT11
DHT dht(DHTPIN, DHTTYPE);

#define DEVICE 1
#define DELAY 3000

void setup()
{
  Heltec.begin(false,true,true,true,BAND);
  Serial.begin(9600);
  LoRa.setSyncWord(0xF3);
  Serial.println("LoRa started");
  dht.begin();
}

void sendTH()
{
  String values = "";
  LoRa.beginPacket();
  float h = dht.readHumidity();
  float t = dht.readTemperature();
  if (isnan(h) || isnan(t)) {
    Serial.println(F("Failed to get data from sensor"));
    return;
  }
  String hS = (String)h;
```

```
    String tS = (String)t;
    String dS = (String)DEVICE;
    values = "{\"t\":"+tS+",\"h\":"+hS+",\"d\":"+dS+"}";
    Serial.println(values);
    LoRa.print(values);
    LoRa.endPacket();
}


void loop()
{
    delay(DELAY);
    sendTH();
}
```

Let's look at the preceding code in more detail:

- `heltec.h`: We import the library to use the ESP32 + LoRa device. With this, you can use the Wi-Fi and LoRa wireless protocol.

- `DHT.h`: We import the library to read data from the DHT11 sensor.

- BAND: We set the band to use to connect the devices. For Europe, you have to use a value of `868E6`, while for America, you have to use `915E6`.

- DHTPIN: This is a constant value that we use to set the PIN that's used to read data in our ESP32 device. In this case, we are using pin 17. Keep in mind that the pin to use has to support digital information.

- DHTTYPE: This defines the type of sensor. The library that we are using supports the DHT11 and DHT12 sensors.

- DEVICE: This is the device number that is sending data. You must change this value every time you upload the code on a device, just to identify each device using a number.

- DELAY: This is the time to wait until sending the next sensor measure data.

- `setup()`: This function does an initial configuration for the Heltec device and sets the network ID for LoRa using `0xF3` – that is, the network ID for our devices. This value must be set between 0 and `0xFF`.

- `sendTH()`: This function captures and sends the sensor data with the LoRa Wi-Fi protocol in the `{"t":26.2,"h":35.5,"d":1}` format, where t is the temperature in Celsius, h is the humidity in percentages, and d is the device number.

- `loop()`: This function runs as a loop and calls `sendTH()` to capture and send sensor data using LoRa.

> **Important Note**
> You can find the source code of the sender and receiver device at `https://github.com/PacktPublishing/Edge-Computing-Systems-with-Kubernetes/tree/main/ch12/code/arduino`.

In summary, first, we set all the constant values to configure how to read data from the DHT11 sensor and send data using LoRa. Then, the setup is called to prepare reading data from the sensor and the initial configuration for LoRa. Finally, `loop()` runs as a loop that calls `sendTH()`, which sends and receives data. Once you have your device with this code, just turn it on to send data. To stop sending data, you must power off the device.

Pay attention to the lines that contain the `Serial.println` command. This command prints information using the serial port. You can troubleshoot what is happening in your device using Arduino when your device is powered with your USB port from your laptop by opening **Tools | Serial Monitor**. Doing this will show all the `Serial.println` outputs inside the window:

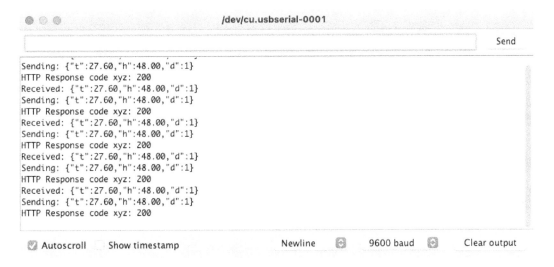

Figure 12.8 – Monitoring with Heltec ESP32 devices in Arduino

> **Important Note**
> To connect your device to macOS, you need a USB A-to-USB C adaptor. If you need a power source with at least 5 volts, you can also use a battery bank instead of connecting the device to a laptop or a computer.

Now that your Heltec sender device is working, you can start uploading the code for the receiver device.

Programming the ESP32 microcontroller to receive sensor data

Now, we have to configure our Heltec ESP32 device to receive the sensor data and send it to the far edge cluster by doing a request across the wireless network. To start, we must create another file by clicking **File | New** and replacing the default content with the following code:

```
#include "heltec.h"
#include "WiFi.h"
#include <HTTPClient.h>
#define BAND     915E6
#define METRICS_IP "192.168.0.241"
void setup()
{
  Heltec.begin(false, true, true, true, BAND);
  Serial.begin(9600);
  LoRa.setSyncWord(0xF3);
  Serial.println("LoRa started");
  WIFISetUp();
}

void WIFISetUp(void)
{
  WiFi.disconnect(true);
  delay(100);
  WiFi.mode(WIFI_STA);
  WiFi.setAutoConnect(true);
  WiFi.begin("NET_NAME","PASSWORD");
  delay(100);

  byte count = 0;
  while(WiFi.status() != WL_CONNECTED && count < 10)
  {
    count ++;
    delay(500);
    Serial.println("Connecting...");
  }
```

```
  if(WiFi.status() == WL_CONNECTED)
    Serial.println("Connected OK");
  else
    Serial.println("Failed");
}

void callURL(String data)
{
  String postData = data;
  Serial.println("Sending: " + postData);
  WiFiClient client;
  HTTPClient http;
  http.begin(client, "http://"+((String)METRICS_IP)+":3000/
device");
  http.addHeader("Content-Type","application/json");
  int httpResponseCode = http.POST(postData);
  Serial.println("HTTP Response code xyz: "+(String)
httpResponseCode);
  http.end();
}

void loop()
{
  onReceive(LoRa.parsePacket());
}

void onReceive(int packetSize)
{
  String incoming = "";
  if (packetSize == 0) return;

  while (LoRa.available())
    incoming += (char)LoRa.read();

  Serial.println("Received: " + incoming);
  callURL(incoming);
}
```

Let's understand the code a little bit:

- `Heltec.h, WiFi.h, HTTPClient.h`: These are the libraries that we are using for the receiver. `Heltec.h` is used to send data with LoRa, `WiFi.h` is the Wi-Fi Heltec library to connect to the wireless network, and `HTTPClient.h` is used to send a request to our far edge server with the sensor data.

- BAND: Here, we set the same band that's used in the sender device.

- METRICS_IP: This is the IP address of the `metrics` service in your cluster. To get this value, go to the *Deploying a service to store sensor data in a MySQL database* section. Replace this value before uploading the code to your device.

- `setup()`: Here, we configure the Heltec device to receive data from the same LoRa network defined by `setSyncWord`. It also configures the Wi-Fi connection.

- `WIFISetUp(void)`: Here, we configure the Wi-Fi connection. To do so, you must replace NET_NAME with your network name and PASSWORD with the necessary password to access your connection.

- `callURL(String data)`: This calls the `metrics` service in your cluster. The URL to access it will be something like `http://METRICS_IP:3000/device`, but this function automatically generates this URL using the value of the `METRICS_IP` constant.

- `onReceive(int packetSize)`: This receives information that's been sent to the configured network using the LoRa protocol and then sends that information to the `metrics` service in the far edge cluster using an HTTP POST request.

- `loop()`: This function runs as a loop and calls `onReceive(int packetSize)`, which gets LoRa packets that contains sensor data. Then, it sends these to the `metrics` endpoint in the far edge cluster.

To summarize, first, we configure the device so that we can connect to the same LoRa network. We must also configure the Wi-Fi, which has access to the far edge server. `loop()` constantly checks if it has received some data to send it to the `metrics` server in the far edge cluster.

Now, upload the code to your device and turn it on, as we did in the *Programming the ESP32 microcontroller to send sensor data* section. With that, our devices have been configured, so let's move to the last step and configure Grafana to show all our data in a dashboard, using the sensor data stored in MySQL.

Visualizing data from ESP32 microcontrollers using MySQL and Grafana

Now, let's finish off our implementation of a real-time temperature and humidity system. For this, we are going to use Grafana to create our reports and MySQL as our source of data to feed the reports. You can deploy this software in Kubernetes in the cloud or a private cloud using a network that can be accessed by your edge clusters. In this section, we are assuming that we are using Kubernetes in the cloud. To start creating our reports, follow these steps:

1. Create the necessary namespace monitoring:

```
$ cat <<EOF | kubectl apply -f -
apiVersion: v1
kind: Namespace
metadata:
  creationTimestamp: null
  name: monitoring
spec: {}
status: {}
EOF
```

2. Create a ConfigMap to create a default data source that contains our MySQL connection:

```
$ cat <<EOF | kubectl apply -f -
apiVersion: v1
kind: ConfigMap
metadata:
  name: grafana-datasources
  namespace: monitoring
  labels:
    grafana_datasource: "true"
data:
  datasource.yaml: |-
    apiVersion: 1
    datasources:
      - name: sensor_data
        type: mysql
        url: mysql.default.svc
        access: proxy
```

```
                    database: sensor_data
                    user: lora_mysql
                    secureJsonData:
                      password: lora123-
                    isDefault: true
        EOF
```

This will be the default data source configured in your `grafana` deployment.

> **Important Note**
> You can use a `Secret` object to secure sensitive data, but we are using `ConfigMap` to simplify this example.

3. Deploy Grafana so that it can use the previous ConfigMap by running the following code:

```
$ cat <<EOF | kubectl apply -f -
apiVersion: apps/v1
kind: Deployment
metadata:
  name: grafana
  namespace: monitoring
spec:
  replicas: 1
  selector:
    matchLabels:
      app: grafana
  template:
    metadata:
      name: grafana
      labels:
        app: grafana
    spec:
      containers:
      - name: grafana
        image: grafana/grafana:8.4.4
        ports:
        - name: grafana
```

```
              containerPort: 3000
          resources:
            limits:
              memory: "1Gi"
              cpu: "1000m"
            requests:
              memory: 500M
              cpu: "500m"
          volumeMounts:
            - mountPath: /var/lib/grafana
              name: grafana-storage
            - mountPath: /etc/grafana/provisioning/
    datasources
              name: grafana-datasources
              readOnly: false
        volumes:
          - name: grafana-storage
            emptyDir: {}
          - name: grafana-datasources
            configMap:
                defaultMode: 420
                name: grafana-datasources
    EOF
```

4.   Let's create the service to access Grafana:

```
$ cat <<EOF | kubectl apply -f -
apiVersion: v1
kind: Service
metadata:
  creationTimestamp: null
  labels:
    app: grafana
  name: grafana
  namespace: monitoring
spec:
  ports:
```

```
    - port: 3000
      protocol: TCP
      targetPort: 3000
  selector:
    app: grafana
  type: ClusterIP
EOF
```

5. Use port-forward to forward the previous Grafana service. This will help us connect to Grafana locally:

```
$ kubectl port-forward svc/grafana 3000 -n monitoring
--address 0.0.0.0
```

6. Go to http://localhost:3000. When the login page appears, use the username admin and password admin and click on the **Log in** button. After that, you will be you for new credentials:

Figure 12.9 – Grafana login

7.   After logging in, you can check that your default data source is set to MySQL by going to **Configuration | Data sources**:

Figure 12.10 – Grafana configuration menu

You will see something similar to the following:

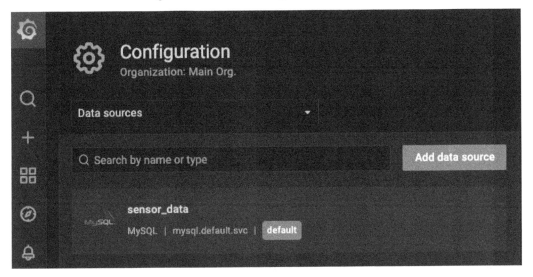

Figure 12.11 – sensor_data default data source

8.   Now, click on **Create | Dashboard**:

Figure 12.12 – Creating a dashboard

Then, click **Add a new panel**:

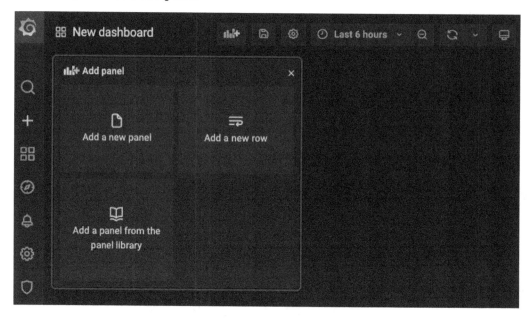

Figure 12.13 – Grafana – The Add panel page

9.   Then, configure the new dashboard with the following query and values:

```
SELECT
    UNIX_TIMESTAMP(time) AS "time",
    temperature_c AS "Temperature(Celcius)",
    temperature_f AS "Temperature(Farenheit)",
    humidity AS "Humidity(%)"
FROM metric
WHERE
    $__timeFilter(time)
    and device=1
ORDER BY time
```

You must edit the default query by clicking on the pencil icon. Then, copy the previous query there:

Figure 12.14 – Editing the default MySQL query

We have set **Title** to **Device 1 Sensor Data** here. The new dashboard will look like this:

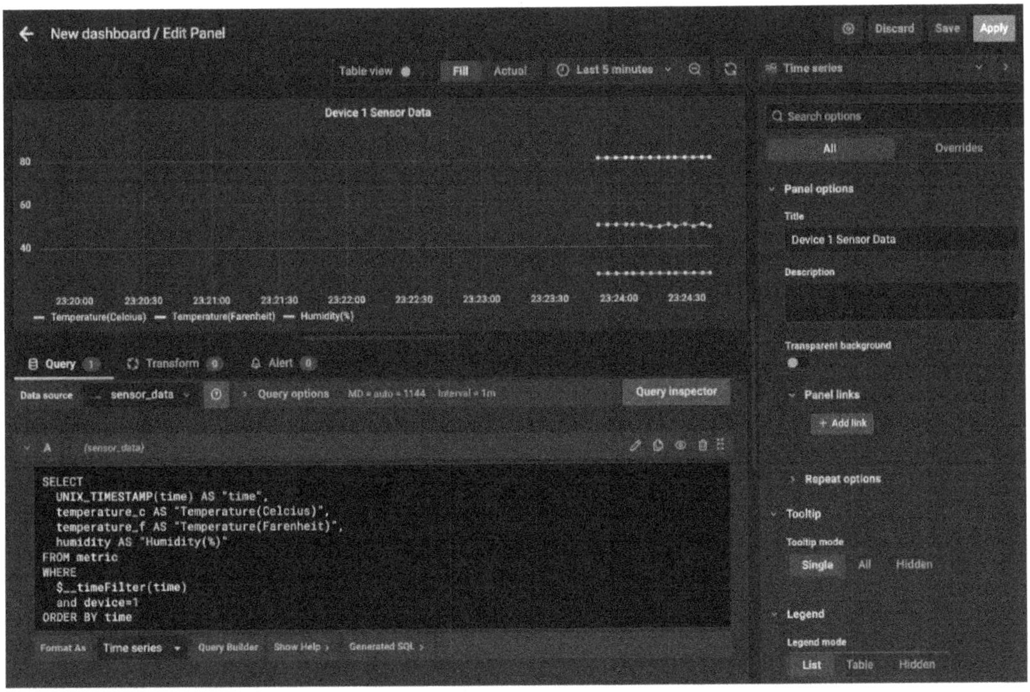

Figure 12.15 – The New dashboard/Edit Panel window

After changing the query if data is available, you will see a graph with three lines – one representing the temperature in Celsius, another representing the temperature in Fahrenheit, and the humidity as a percentage. Remember to set the dashboard to visualize the proper range of data – for example, to show data from the last 5 minutes – and refresh the dashboard every 5 seconds.

10. After that, save your dashboard by clicking on the **Save** button.

Important Note

Check out *Chapter 11*, *Monitoring the Edge with Prometheus and Grafana*, for more details about customizing your dashboard.

11. Finally, your dashboard will look as follows:

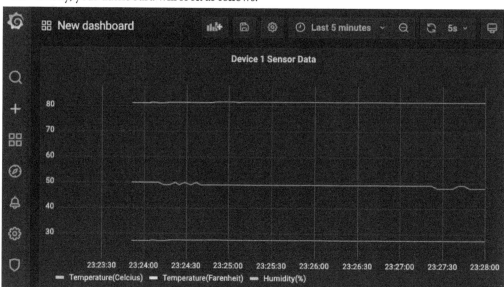

Figure 12.16 – ESP32 monitoring dashboard in Grafana

At this point, you are visualizing the data from your Heltec devices to send and receive data. This device interacts with your far edge cluster. If you chose to deploy Grafana and MySQL on the cloud, this scenario is also interacting with the cloud layer. All these components interact with each other to do their job. Remember that this is a simple implementation that you can extend to your own. Now, let's summarize what we have learned in this chapter.

Summary

In this chapter, we explored how to implement an edge computing system by using LoRa devices to send and receive sensor data. Finally, we implemented a dashboard using MySQL and Grafana. In this way, the LoRa wireless protocol represents a way to implement lower-cost systems that need to transmit information close to the edge. Therefore, LoRa is a common choice as a transmission protocol for edge devices and IoT applications. In the next chapter, we are going to use a GPS module to extend the range of communication and databases to implement geolocation applications.

Questions

Here are a few questions to validate your new knowledge:

- What are the uses and advantages of using the LoRa wireless protocol?
- What is LoRaWAN?

- How can I use a Heltec ESP32 + LoRa device to send sensor data?

- How can I use Arduino IDE to program ESP32 devices?

- How can I create a simple gateway to send data coming from LoRaWAN to LAN?

- How can I use MySQL and Grafana to create reports?

Further reading

Please refer to the following references for more information on the topics covered in this chapter:

- What is the LoRaWAN specification?: `https://lora-alliance.org/about-lorawan`

- What is LoRa?: `https://www.semtech.com/lora/what-is-lora`

- LoRa applications: `https://www.semtech.com/lora/lora-applications`

- LoRaWAN Frequency Plans: `https://www.thethingsnetwork.org/docs/lorawan/frequency-plans`.

- CP210x USB to UART Bridge VCP Drivers: `https://www.silabs.com/developers/usb-to-uart-bridge-vcp-drivers`

- Arduino download software: `https://www.arduino.cc/en/software`

- HttpClient documentation and examples: `https://github.com/amcewen/HttpClient`

- All fixes to run ESP32/Arduino on Arduino IDE and Platform I/O using MacOS Big Sur and Newer: `https://www.youtube.com/watch?v=zkyoghpT8_U`

- ESP32 + LoRa Heltec documentation: `https://heltec-automation-docs.readthedocs.io/en/latest/esp32/index.html`

- Heltec Automation Docs Page: `https://heltec-automation-docs.readthedocs.io/en/latest`

- Heltec ESP32 LoRaWAN library and examples: `https://github.com/HelTecAutomation/ESP32_LoRaWAN`

- WIFI LoRa 32(V2) Pinout Diagram: `http://resource.heltec.cn/download/WiFi_LoRa_32/WIFI_LoRa_32_V2.pdf`

- Provisioning Grafana: `https://github.com/grafana/grafana/blob/main/docs/sources/administration/provisioning/index.md`

13
Geolocalization Applications Using GPS, NoSQL, and K3s Clusters

One of the growing use cases for edge computing is the implementation of a system for tracking cargos and logistics. Sometimes, this tracking involves monitoring and getting metrics that can be used to optimize packages' delivery times, reduce gas consumption, and so on. One of the important technologies that you can use for this is the **Global Positioning System** (**GPS**). GPS can help you to obtain the coordinates of an object when it is moving in real time. This, together with Kubernetes at the edge, results in a powerful combination of technologies to create geolocalization systems, also called geo-tracking systems.

In this chapter, we're going to cover the following main topics:

- Understanding how GPS is used in a geo-tracking system
- Using Redis to store GPS coordinates data
- Using MongoDB to store your devices' tracking data
- Creating services to monitor your devices in real time using GPS
- Configuring your Raspberry Pi to track your device using GPS
- Visualizing your devices using Leaflet library in real time
- Deploying a real-time map and report application to track your devices

Technical requirements

To deploy our databases in this chapter, you need the following:

- A single node K3s cluster using an ARM device. In this case, we are going to use the Raspberry Pi 4B model with 8 GB. We are going to use the Raspberry Pi OS lite (64-bit) operating system with no desktop environment.

- Multiple VK-162 G-Mouse USB GPS dongle navigation modules for your edge Raspberry devices.

- A battery bank and a USB 2.0 A-Male to USB C cable. You can also power your Raspberry Pi with your car's USB charger port.

- A Kubernetes cluster hosted in your public cloud provider (AWS, Azure, or GCP) or your private cloud.

- Basic knowledge of programming, especially Python and JavaScript.

- Clone the repository at `https://github.com/PacktPublishing/Edge-Computing-Systems-with-Kubernetes/tree/main/ch13` if you want to run the YAML configuration by using `kubectl apply` instead of copying the code from the book. Take a look at the `python` directory inside the `code` directory and the `yaml` directory for YAML configurations inside the `ch13` directory.

With this, you can start to implement your geolocation system using edge computing. Let's start to understand how GPS works in our first section.

Understanding how GPS is used in a geo-tracking system

For this chapter, our goal will be to build a geolocation system, also called a geo-tracking system. This means that we are going to build a system that gets GPS coordinates or positions from vehicles. In our use case, we are assuming that our vehicles will be used to deliver packages. Our vehicles will be equipped with a Raspberry Pi device and a GPS module. This hardware will collect GPS coordinates, using latitude and longitude to send them to the cloud. Then, our application will show the live positions of all the vehicles and a report to show the route of vehicles within a date range. In general, these are the main features that our geo-tracking system will have:

- A real-time map showing the position of all delivery vehicles

- A map showing the nearby delivery stops of each vehicle

- A report that shows delivery routes for a vehicle between a date range

This geo-tracking system is represented using the following diagram:

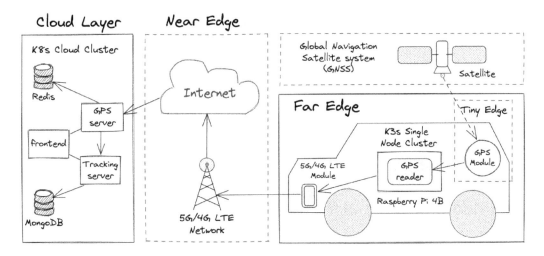

Figure 13.1 – A geolocation application edge diagram

Now, let's explain how this geo-tracking system is implemented by describing the edge computing layers:

- **Cloud layer**: Here, we are going to install a Kubernetes cluster in our preferred cloud provider. Inside this cluster, we are going to install three main applications. The GPS server will receive requests from the vehicles. It will also save the collected positions in Redis for the real-time map and the logs in MongoDB for the report that shows delivery routes. Finally, we will have the frontend application that contains the web application to show the real-time map and the report.

- **Near edge**: This layer represents all the information that will move from the far edge to the near edge using the LTE network. This means that all the GPS information will be sent across the internet to its final destination, the cloud layer.

- **Far edge**: Here, we are going to find the vehicles equipped with a Raspberry Pi; this device will use a GPS module and an internet connection to send the GPS coordinates. Our Raspberry Pi will have K3s installed. Inside this K3s single-node cluster, we are going to find the GPS reader. This application is going to read GPS information and send it to the cloud. However, you can also include additional applications to add more functionalities – for example, showing an OLED screen with GPS information or other processed data, such as velocity, using the GPS coordinates to calculate it. Hence, this part represents the local process at the edge.

  To connect the Raspberry Pi to the internet, you can use a 5G or 4G LTE module or your smartphone, which already includes this kind of module. To simplify the example, we are going to use the access point from a smartphone to share the internet with the Raspberry Pi device.

- **Tiny edge**: Here, we are going to find the GPS module that our edge device will use to get GPS coordinates. Our GPS module is going to use the **Global Navigation Satellite System** (**GNSS**), which is a global system of satellites that gives you GPS coordinates. This will be the main data used in our implementation. You can also use an LTE 5G/4G module with GPS integrated to speed up GPS module initialization to capture GPS coordinates, but this could be expensive compared with the VK-162 G-Mouse USB module. In this case, we are going to use the VK-162 module to simplify the implementation and reduce costs for this prototype implementation.

In summary, our vehicle on the far edge is going to read information from the GPS module on the tiny edge. After reading the information and doing some processing, the information will be sent to the cloud layer using the near edge. Once all the information is received, it will be stored in Redis and Mongo to show the real-time map and the report using the frontend application.

Using Redis to store GPS coordinates data

As we explained in *Chapter 10, SQL and NoSQL Databases at the Edge*, Redis is a key-value database that is pretty lightweight when using resources. Redis exclusively uses RAM memory to store its data but can persist when using snapshot configuration, which basically stores this data on the disk. Redis can also store geolocation data, storing GPS coordinates and tuples with latitude and longitude values. Redis stores this information with the field's latitude, longitude, and a name. Redis also calls this data a **geospacial index**. Redis also includes the ability to return coordinates close to a circular area with this type of data. In this use case, Redis will be used to calculate all this information. For this specific use case, we are going to use the GEOADD and GEOSEARCH commands to implement our geolocalization application. But first, let's install Redis in the cloud to store some geolocation data. For this, follow these steps:

1. First, let's create a **PersistentVolumeClaim** for Redis to persist data:

```
$ cat <<EOF | kubectl apply -f -
apiVersion: v1
kind: PersistentVolumeClaim
metadata:
  name: db-pv-claim-1
spec:
  accessModes:
    - ReadWriteOnce
  resources:
    requests:
      storage: 5Gi
EOF
```

2. Now, create a ConfigMap to configure Redis to use an authentication password:

```
$ cat <<EOF | kubectl apply -f -
apiVersion: v1
kind: ConfigMap
metadata:
  name: redis-configmap
data:
  redis-config: |
    dir /data
    requirepass YOUR_PASSWORD
EOF
```

3. Create the deployment for Redis using the previous **ConfigMap** called `redis-configmap` and mounted as the `redis.conf` file. We also use the **PersistentVolumeClaim** called `db-pv-claim-1`, and some resource limits for the deployment setting the CPU and memory. Let's create the deployment by running the following command:

```
$ cat <<EOF | kubectl apply -f -
apiVersion: apps/v1
kind: Deployment
metadata:
  creationTimestamp: null
  labels:
    app: redis
  name: redis
spec:
  replicas: 1
  selector:
    matchLabels:
      app: redis
  strategy: {}
  template:
    metadata:
      creationTimestamp: null
      labels:
        app: redis
    spec:
```

```
      containers:
      - name: redis
        image: redis:6.2
        command:
          - redis-server
          - /redisconf/redis.conf
        ports:
        - containerPort: 6379
        resources:
          limits:
            cpu: "0.2"
            memory: "128Mi"
        volumeMounts:
        - mountPath: "/data"
          name: redis-storage
        - mountPath: /redisconf
          name: config
      volumes:
        - name: config
          configMap:
            name: redis-configmap
            items:
            - key: redis-config
              path: redis.conf
        - name: redis-storage
          persistentVolumeClaim:
            claimName: db-pv-claim-1
  status: {}
  EOF
```

This time, we are not going to use an image for ARM 64 bits.

4. Now, create the service for Redis by opening port 6379:

```
$ cat <<EOF | kubectl apply -f -
apiVersion: v1
kind: Service
metadata:
```

```
    labels:
      app: redis
    name: redis
  spec:
    ports:
    - port: 6379
      protocol: TCP
      targetPort: 6379
    selector:
      app: redis
    type: ClusterIP
EOF
```

Now, we have Redis installed. Let's move to install Mongo to store log information with this data.

Using MongoDB to store your device's tracking data

MongoDB is a document-oriented NoSQL database that stores the information using JSON format. It also has the capability to store location data. In this use case, we are going to use MongoDB to store our geolocation data; this means storing all coordinates (latitude and longitude) that the GPS captures on the devices for later reports. MongoDB can perform some special manipulation for geolocation data, but in this case, we will use it just to store data in JSON format. To install MongoDB in the cloud, follow the next steps:

1. Create a **PersistentVolumeClaim** for MongoDB, to persist data:

```
$ cat <<EOF | kubectl apply -f -
apiVersion: v1
kind: PersistentVolumeClaim
metadata:
  name: db-pv-claim-2
spec:
  accessModes:
    - ReadWriteOnce
  #storageClassName: your_driver
  resources:
    requests:
      storage: 5Gi
EOF
```

> **Important Note**
>
> You can change the storage class if you install Longhorn or another storage driver, or if you are using the storage class provided by your cloud provider. Just uncomment the `storageClassName` line by removing the # character.

2. Deploy your custom configuration to enable clients to connect to MongoDB:

```
$ cat <<EOF | kubectl apply -f -
apiVersion: v1
kind: ConfigMap
metadata:
  name: mongo-configmap
data:
  mongod-conf: |
    dbpath=/var/lib/mongodb
    logpath=/var/log/mongodb/mongodb.log
    logappend=true
    bind_ip = 0.0.0.0
    port = 27017
    journal=true
    auth = true
EOF
```

This exposes MongoDB to listening in port 27017 across the network.

3. Create the deployment using the **ConfigMap** called `mongo-configmap`, our **PersistentVolumeClaim**, and the MONGO_INITDB_ROOT_USERNAME, MONGO_INITDB_ROOT_PASSWORD, and MONGO_INITDB_DATABASE variables that set the initial root username, an additional user to connect and their passwords to be used when connecting to MongoDB:

```
$ cat <<EOF | kubectl apply -f -
apiVersion: apps/v1
kind: Deployment
metadata:
  labels:
    app: mongo
  name: mongo
spec:
  replicas: 1
```

```yaml
selector:
  matchLabels:
    app: mongo
template:
  metadata:
    labels:
      app: mongo
  spec:
    containers:
    - name: mongo
      image: mongo:4.4
      env:
      - name: MONGO_INITDB_ROOT_USERNAME
        value: "admin"
      - name: MONGO_INITDB_ROOT_PASSWORD
        value: "YOUR_PASSWORD"
      - name: MONGO_INITDB_DATABASE
        value: "mydatabase"
      ports:
      - containerPort: 27017
      resources:
        limits:
          cpu: "0.2"
          memory: "200Mi"
      volumeMounts:
      - mountPath: "/data/db"
        name: mongo-storage
      - mountPath: /mongoconf
        name: config
    volumes:
      - name: config
        configMap:
          name: mongo-configmap
          items:
          - key: mongod-conf
```

```
                        path: mongod.conf
                  - name: mongo-storage
                    persistentVolumeClaim:
                        claimName: db-pv-claim-2
         EOF
```

> **Important Note**
>
> We are using some values directly to configure the deployment to simplify the example. But it's a best practice to use secrets to protect sensitive data. You can explore *Chapter 10, SQL and NoSQL Databases at the Edge*, for more examples. We are also using version 4.4 in case you want to install MongoDB on an ARM device.

4. Now, create the service that exposes your MongoDB deployment as a service accessible inside the cluster (MongoDB uses port 27017 to connect):

```
$ cat <<EOF | kubectl apply -f -
apiVersion: v1
kind: Service
metadata:
  labels:
    app: mongo
  name: mongo
spec:
  ports:
  - port: 27017
    protocol: TCP
    targetPort: 27017
  selector:
    app: mongo
  type: ClusterIP
EOF
```

Now, your MongoDB database has been installed. So, let's deploy our GPS server application that will store data in Redis and our MongoDB in the next section.

Creating services to monitor your devices in real time using GPS

In our use case, we are going to deploy a service that sends data from our edge device after some processing in the cloud. The goal of this use case is to have a global geolocation system for multiple vehicles delivering packages, showing their location in real time. For this, we are going to create a `gps-server` deployment that stores all the coordinates for our units in Redis and Mongo. We are going to use the Python Flask library to create this service. Let's explore the main sections of the following pseudocode mixed with Python:

```
<imported libraries>
<app_initialization>
<CORS configuration>

def redisCon():
<return Redis connection object>

@app.route("/client/<cid>/position", methods=["POST"])
def setPosition(cid):
   <Call redisCon>
   <Store of data in a Redis hash data type using
    the fields cid,lat,lng
    in the hash key named client:{cid}:position>
   <set the expiration of the key>
   <call the tracking-server in /client/{cid}/position
    to store the position in Mongo>
   return  {"client_id":cid,"setPosition":"done"}

@app.route("/clients/positions/unit/<unit>/r/<radius>"
            ,methods=["GET"])
def getPositions(unit,radius):
   <Call redisCon>
   <Search for client:*:position keys>
      <Search the near geospacial index for
       the current position>
      <Add the position to data Array>
```

```
        <Returns the near positions for each unit in JSON>
        return jsonify({"clients":data})


@app.route("/client/<cid>/stops", methods=["POST"])
def setStops(cid):
        <Call redisCon>
        <GET json values stops to set>
        <Store the stops in the key client:{cid}:stops >
        return jsonify({"setStops":"done"})


<App initialization in port 3000>
```

Let's focus on the following functions:

- **redisCon**: This function sets the Redis connection. This application is going to use the Redis service created in the *Using Redis to store GPS coordinates data* section.

- **setPosition**: Each time our application receives the `/client/<cid>/position` URL, the function will get the `<cid>` value that represents a connected client that sends information to this service – in this case, our delivery vehicles. Every time the information is received, it is stored in the key with the form `client:{cid}:position` inside Redis, and stores the latitude as a `lat` variable, the longitude as `lng`, and the client ID or vehicle number as `cid`. It also sets an expiration time as 180 seconds or 3 minutes. After calling the tracking server to store this coordinate in MongoDB, it returns the following JSON response: `{"client_id":cid,"setPosition":"done"}`.

- **getPositions**: Each time our application receives the `/clients/positions/unit/<unit>/r/<radius>` URL, the function connects to Redis and gets all the keys with the form `client:<cid>:position`, which contains the current GPS position of each vehicle. Then gets the near stops to this position, using the Redis command `geosearch`. The returned JSON will look like: `{"clients":[{"cid":1,"lat":0.0,"lng":0.0,"near":["stop1"]}]}`.

- **setStops**: Each time our application receives the `/client/<cid>/stops` URL, the function will get `<cid>` and store all the positions as a geospatial index in the key with the `client:{cid}:stops` form. Inside this key, each position will be stored with the name sent as part of the JSON data using `curl`. These stops are stored for 10 hours by default because the stops have to be completed during a workday. These stops will be near to the vehicle with the `cid` number.

After understanding the code, let's deploy our GPS server application in the next section.

Deploying gps-server to store GPS coordinates

The `gps-server` application will receive the GPS coordinates from your edge devices. For this, we have to deploy it and expose it using a load balancer. To deploy the `gps-server` application, follow the following steps:

1. Create the deployment for the GPS server:

```
$ cat <<EOF | kubectl apply -f -
apiVersion: apps/v1
kind: Deployment
metadata:
  creationTimestamp: null
  labels:
    app: gps-server
  name: gps-server
spec:
  replicas: 1
  selector:
    matchLabels:
      app: gps-server
  strategy: {}
  template:
    metadata:
      creationTimestamp: null
      labels:
        app: gps-server
    spec:
      containers:
      - image: sergioarmgpl/gps_server
        name: gps-server
        imagePullPolicy: Always
        env:
        - name: REDIS_HOST
          value: "redis"
        - name: REDIS_AUTH
          value: "YOUR_PASSWORD"
```

```
      - name: ENDPOINT
        value: "http://tracking-server:3000"
    resources: {}
  status: {}
  EOF
```

This deployment uses the following variables:

- **REDIS_HOST**: This is the name of the Redis service; this variable can be customized to fit your needs.

- **REDIS_AUTH**: This is the password to connect to the Redis service.

- **ENDPOINT**: This is the URL of `tracking-server` – in this case, the URL matches the internal `tracking-server` service in port `3000`.

> **Important Note**
>
> To check the code and create your own container, refer to this link: `https://github.com/sergioarmgpl/containers/tree/main/gps-server/src`.

2. Create the service as a LoadBalancer; this IP address will be used in our GPS reader services for each unit or truck:

```
$ cat <<EOF | kubectl apply -f -
apiVersion: v1
kind: Service
metadata:
  creationTimestamp: null
  labels:
    app: gps-server
  name: gps-server-lb
spec:
  ports:
  - port: 3000
    protocol: TCP
    targetPort: 3000
  selector:
    app: gps-server
  type: LoadBalancer
```

```
status:
  loadBalancer: {}
EOF
```

3. Get the load balancer IP for our `gps-server` deployment with the following command:

```
$ GPS_SERVER_IP="$(kubectl get svc gps-server-lb
-o=jsonpath='{.status.loadBalancer.ingress[0].ip}')"
```

You can see the value of the GPS_SERVER_IP environment variable by running the following:

```
$ echo $GPS_SERVER_IP
```

Note that it takes some time after the IP address of the load balancer is provisioned. You can check the state of the services by running the following:

```
$ kubectl get svc gps-server-lb
```

Wait until EXTERNAL_IP is provisioned. Also, note the $GPS_SERVER_IP value, which will be used to configure the gps-reader application on each edge device.

Now, you can set the stops for the first vehicle, represented with value 1. For this, follow the next steps:

1. Use curl to store the stops:

```
$ curl -X POST -H "Accept: application/json" \
-H "Content-Type: application/json" \
--data '{
    "stops":[
    {"name":"stop1","lat":1.633518,"lng": -90.591706},
    {"name":"stop2","lat":2.631566,"lng": -91.591529},
    {"name":"stop3","lat":3.635043,"lng": -92.589982}
    ]
}' http://$GPS_SERVER_IP:3000/client/1/stops
```

2. This will return the following:

```
{"setStops":"done"}
```

Now, we have the gps-server application deployed and exposed using a load balancer. Let's deploy our tracking-server, the one that stores logs about the received GPS positions.

Creating a service to log GPS positions and enable real-time tracking for your devices

Our tracking-server application will be in charge of logging all the received coordinates for each vehicle. This information will be used to show the route of a vehicle in the desired time range using the frontend application. Before deploying tracking-server, let's understand the code of this application:

```
<Imported libraries>
<Application initialization>
<CORS configuration>

def mongoCon():
    <return Mongo connection with tracking collection set>

@app.route("/client/<cid>/position", methods=["POST"])
def storePosition(cid):
    <Get the position JSON values to store it
     in the tracking collection>
    <Get current time and store it using UTC>
    <Call MongoCon function>
    <Store data in the format:
     {"cid":XX,"lat":XX,"lng":XX,"ts":XXXXXXX,"dtxt":XXXXXX}
     Inside the tracking collection in the database
     called mydatabase>
    <return JSON {"client_id":cid,"positionStored":"done"}>

@app.route("/client/<cid>/positions/s/<sdate>/e/<edate>"
           ,methods=["GET"])
def getPositions(cid,sdate,edate):
    <Get the start date to query in the format
     dd-mm-yy-HH:MM:SS and convert it into UTC>
    <get the end date to query in the format
     dd-mm-yy-HH:MM:SS and convert it into UTC>
    <Call MongoCon function>
    <Query the tracking collection to get the
     tracking data for a unit
     or truck between the time range>
    <Return the positions in an array called data>
```

```
return jsonify({"tracking":data})
```

```
<App initialization in port 3000>
```

In this code we can find the following functions:

- **MongoCon**: This function connects to MongoDB and returns a MongoDB object connection, with the collection set to the `tracking` value.

- **storePosition**: Each time our application receives a `POST` request in the `/client/<cid>/position` URL, the function will store the received GPS position in the `{"cid":1,"lat":0.0,"lng":0.0,"ts":166666666,"dtxt":"01-01-22-23:59:59"}` format. `cid` represents the client ID or the number of the vehicle, `lat` and `lng` are used to store the GPS position, `ts` represents the timestamp generated when the coordinate was received, and `dtxt` is the date in text format to reduce transformation time from the timestamp format to the UNIX date format. Once this data is stored in the database, `mydatabase` returns the next JSON: `{"client_id":cid,"positionStored":"done"}`.

- **getPositions**: Each time our application receives a `GET` request in the URL with the `/client/<cid>/positions/s/<sdate>/e/<edate>` form, it returns a JSON response with all the GPS positions between the starting date, `sdate`, and the ending date, `edate`. For this, `tracking-server` connects to Mongo and returns the result of performing this query in this time range. The information will be returned in the following format:

```
{
    "tracking":[
        {"lat":0.0,"lng":0.0,"ts":166666666
        ,"dtxt":"01-01-22-23:59:59"}
    ]
}
```

Now we know how the `tracking-server` application works. Let's deploy this application in the next section.

Deploying tracking-server to store logs from GPS coordinates to be used for vehicles routing report

Our `tracking-server` will be used in our `frontend` application to show the route of a vehicle within the desired time range. Let's deploy our application with the following steps:

1. Deploy `tracking-server` by running the following:

```
$ cat <<EOF | kubectl apply -f -
apiVersion: apps/v1
```

```
kind: Deployment
metadata:
  creationTimestamp: null
  labels:
    app: tracking-server
  name: tracking-server
spec:
  replicas: 1
  selector:
    matchLabels:
      app: tracking-server
  strategy: {}
  template:
    metadata:
      creationTimestamp: null
      labels:
        app: tracking-server
    spec:
      containers:
      - image: sergioarmgpl/tracking_server
        name: tracking-server
        imagePullPolicy: Always
        env:
        - name: MONGO_URI
          value: "mongodb://admin:YOUR_PASSWORD@mongo/
mydatabase?authSource=admin"
        - name: MONGO_DB
          value: "mydatabase"
        - name: TIMEZONE
          value: "America/Guatemala"
        resources: {}
status: {}
EOF
```

This deployment uses the following environment variables:

- **MONGO_URI**: This is the full URI that contains a string to be used to authenticate in MongoDB. It has the `mongodb://USER:PASWORD@HOST/DATABASE?authSource=admin` format. You can customize these credentials and store `MONGO_URI` as a secret.

- **MONGO_DB**: This is the database created in MongoDB to store the tracking collection.

- **TIMEZONE**: This is the time zone used to get the time when the GPS coordinate is stored in the `tracking` collection of MongoDB. Note that our Python code uses the `pytz` library and the ISO 3166 convention for country names. Check the *Further reading* section for more information to set your country's time zone correctly. In this case, we set the country to `America/Guatemala`.

> **Important Note**
>
> You can find out more about the URI on the following page: `https://www.mongodb.com/docs/manual/reference/connection-string`. Remember that we are using hardcoded values to simplify the example, but it's best practice to use secrets. Check out *Chapter 10, SQL and NoSQL Databases at the Edge*, for more details. To check out the code and create your own version of `tracking-server`, refer to the following link: `https://github.com/sergioarmgpl/containers/tree/main/tracking-server/src`.

2. Create a service as a ClusterIP for `tracking-server` to call it inside `gps-server`:

```
$ cat <<EOF | kubectl apply -f -
apiVersion: v1
kind: Service
metadata:
  creationTimestamp: null
  labels:
    app: tracking-server
  name: tracking-server
spec:
  ports:
  - port: 3000
    protocol: TCP
    targetPort: 3000
  selector:
    app: tracking-server
  type: ClusterIP
status:
```

```
      loadBalancer: {}
   EOF
```

3. Create a service as a LoadBalancer for `tracking-server` to call it in our viewer application, which is accessible over the internet:

```
$ cat <<EOF | kubectl apply -f -
apiVersion: v1
kind: Service
metadata:
  creationTimestamp: null
  labels:
    app: tracking-server
  name: tracking-server-lb
spec:
  ports:
  - port: 3000
    protocol: TCP
    targetPort: 3000
  selector:
    app: tracking-server
  type: LoadBalancer
status:
  loadBalancer: {}
EOF
```

4. Get the load balancer IP of our `tracking-server` deployment with the following command:

```
$ TRACKING_SERVER_IP="$(kubectl get svc tracking-
server-lb -o=jsonpath='{.status.loadBalancer.ingress[0].
ip}')"
```

You can see the value of the TRACKING_SERVER_IP environment variable by running the following:

```
$ echo $TRACKING_SERVER_IP
```

The `tracking-server` application has been deployed. Now, let's configure our device to run our reader application.

Configuring your Raspberry Pi to track your device using GPS

Before using the GPS module on your Raspberry Pi, you have to follow the following steps:

1. Install the Raspberry Pi OS Lite (64-bit) on your device; you can check out *Chapter 2, K3s Installation and Configuration*, for more details.

2. Log in to your device and set an initial user name and password.

3. Run `raspi-config` with the following command:

   ```
   $ sudo raspi-config
   ```

   You will see a screen like the following:

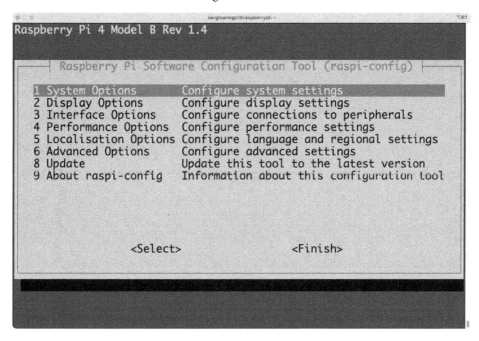

Figure 13.2 – The raspi-config main menu

4. To configure the wireless network, go to the **System Options | Wireless LAN** menu.

5. You will see a **Choose the country where your Raspberry will be used** message, and then click **Ok**.

6. After that, the **Wireless LAN country** message will appear. Select your country and then click **Ok**.

7. The **Please Enter SSID** message will appear. Click **Ok** and press **Enter**.

8. Now, the **Please Enter passphrase** message will appear. Click **Ok** and press *Enter*.

9. Upon returning to the main menu, select **Finish** and press *Enter* to exit.

10. Activate the SSH, choosing the **Interface Options** | **SSH** menu. This will show the **Would you like the SSH server to be enabled?** message. Choose **Yes** and press *Enter*. After that, the **The SSH server is enabled** message will appear.

11. To check the IP of your Raspberry Pi, run the following command:

```
$ ifconfig -a
```

The output will look like the following:

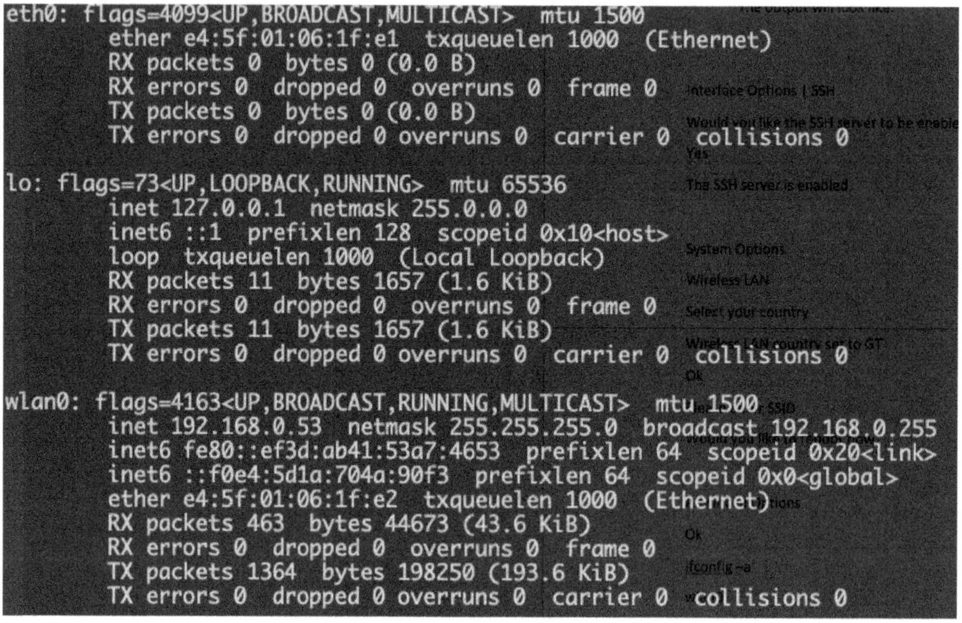

Figure 13.3 – The ifconfig output

Take note of the IP address next to the word `inet` word in the `wlan0` network interface; this will be the IP address of your Raspberry Pi.

12. Log in to your device using the previous IP address found using SSH:

```
$ ssh YOUR_USER@RASPBERRY_IP
```

13. Add the next kernel parameters to enable the use of container by adding these values in the `/boot/cmdline.txt` file; remember that you need root permissions to modify this file:

```
cgroup_memory=1 cgroup_enable=memory
```

14. Connect your VK-162 G-Mouse GPS module to one of the USB ports of your Raspberry; after some seconds, the /dev/ttyACM0 device will be ready to be used.

15. Restart your device to apply these changes:

```
$ sudo shutdown -r now
```

16. (*Optional*) If you want to configure other features, log in to your device and run the following:

```
$ sudo raspi-config
```

17. (*Optional*) Activate the **Inter-Integrated Circuit** (**I2C**) support in your device – for example, to connect an OLED screen, go to the **Interface Options** | **I2C** menu after running raspi-config.

18. (*Optional*) Then, a dialog will show **Would you like the ARM I2C interface to be enabled?**. Select **Yes** and press *Enter*.

19. (*Optional*) After the previous dialog, the **The ARM I2C interface is enabled** message will appear. Press *Enter* to choose the **Ok** button.

20. (*Optional*) Upon returning to the main menu, select **Finish** and press *Enter* to exit.

21. To finish, let's install K3s by running the following:

```
$ curl -sfL https://get.k3s.io | INSTALL_K3S_EXEC="--
write-kubeconfig-mode 644" sh -s -
```

22. Run the following command to see whether your K3s single-node cluster is running:

```
$ kubectl get nodes
```

This will show something like the following:

```
NAME          STATUS    ROLES                AGE    VERSION
raspberrypi   Ready     control-plane,master 53s    v1.23.6+k3s1
```

Figure 13.4 – The kubectl get nodes output

> **Important Note**
>
> You can find out more about how to use raspi-config at the following link: https://geek-university.com/raspi-config.

Now, you have your Raspberry Pi installed with Raspberry Pi OS Lite, which is ready to be used together with your GPS module. In the next section, let's move to deploy the GPS reader application.

Understanding the GPS reader code to send GPS coordinates

Now, the only remaining part is to install gps-reader in the K3s single-node cluster installed on your Raspberry. This application will run as a container using the Kubernetes Pods. However, before installing our gps-reader application, let's understand the code first:

```
<Imported libraries>

while True:
    <Set serial Device /dev/ttyACM0 with baud rate 9600>
    ser=serial.Serial(device, baudrate=9600, timeout=0.5)
    <Set the PynMEA2 reader>
    <Read data from the device>

    <Read for GRPMC lines>
     <Extract latitude, longitude>
     <Call /client/{cid}/position from GPS Server
      To store the position in Redis>
     <If cannot read data show
      "No GPS data to send">
```

The previous code contains an infinite loop that reads the output of the /dev/ttyACM0 device every half second. Our VK-162 G-Mouse GPS module uses the **National Marine Electronics Association (NMEA)** specification to represent GPS coordinates. The previous code scans the output to look for the GRPMC field to get the latitude and longitude coordinates using the PynMEA2 library. Once the library extracts the coordinates, it calls the GPS server endpoint to store the current GPS position of the vehicle in Redis and log it in MongoDB.

Be aware that the GPS module delays a little bit after the module starts receiving a GPS coordinate. It could take some minutes before the GPS module starts to receive GPS coordinates.

To see what your device is doing, run the cat /dev/ttyACM0 command. If the module is not receiving coordinates yet, it will show something like the following:

```
$GPRMC,052326.00,V,,,,,,,,,,N*7D
$GPVTG,,,,,,,,,N*30
$GPGGA,052326.00,,,,,0,00,99.99,,,,,,*66
$GPGSA,A,1,,,,,,,,,,,,,99.99,99.99,99.99*30
$GPTXT,01,01,01,NMEA unknown msg*58
$GPTXT,01,01,01,NMEA unknown msg*58
$GPGSV,1,1,02,01,,,30,22,,,36*7C
$GPGLL,,,,,052326.00,V,N*4A
```

> **Important Note**
>
> The GPRMC or GPGLL fields are empty in some parts when the module is not receiving coordinates. These missing values contain the latitude and longitude obtained by the GPS module.

When the device starts to receive data, you will see something like the following:

```
$GPRMC,054003.00,A,1437.91511,N,09035.52679,W,0.077,,020622,,,
D*6D
$GPVTG,,T,,M,0.077,N,0.142,K,D*21
$GPGGA,054003.00,1437.91511,N,09035.52679,W,2,06,2.54,1668.7
,M,-4.9,M,,0000*68
$GPGSA,A,3,22,01,48,31,32,21,,,,,,,3.95,2.54,3.02*02
$GPTXT,01,01,01,NMEA unknown msg*58
$GPGSV,4,1,13,01,18,301,33,10,49,124,11,16,20,189,12,21,29,276,
24*74
$GPG
SV,4,2,13,22,39,008,32,23,18,135,,25,19,052,11,26,47,169,09*79
$GPGSV,4,3,13,27,02,204,18,31,64,342,30,32,35,037,29,46,43,252
,*7A
$GPGSV,4,4,13,48,47,250,30*40
$GPGLL,1437.91511,N,09035.52679,W,054003.00,A,D*70
```

The GPGLL line contains all the information about latitude and longitude that we are looking for.

> **Important Note**
>
> Depending on your device configuration, the GPRMC line can include the elevation data. In the previous output, the elevation information is not configured, so the line will not include this information, but the device can be configured to get the elevation information too.

Now, we know how our application reads information from the GPS module. Let's deploy our application in our device with K3s installed.

Deploying gps-reader to send GPS coordinates to the cloud

One advantage of using K3s is that if your application is complex, you can deploy your application separated as modules or microservices, and you can update these pieces without affecting the others. In this case, we are only using one piece called gps-reader. This application reads the GPS module from the device using a Pod. In this case, we are using a configuration that enables us to read the / dev folder from the device with just the necessary permissions to access the /dev/ttyACM0 device, where the GPS module shows the GPS coordinates. This device can change, depending on the GPS module that you are using.

To create a reader on your device, run the following steps:

1. Create the gps-reader Pod to start reading GPS coordinates from your module:

```
$ cat <<EOF | kubectl apply -f -
apiVersion: v1
kind: Pod
metadata:
  name: gps-reader
spec:
  containers:
  - image: sergioarmgpl/gps_reader
    name: gpsreader
    imagePullPolicy: Always
    env:
    - name: DEVICE
      value: "/dev/ttyACM0"
    - name: CLIENT_ID
      value: "1"
    - name: ENDPOINT
      value: "http://<GPS_SERVER_IP>:3000"
    securityContext:
      privileged: true
      capabilities:
        add: ["SYS_ADMIN"]
    volumeMounts:
    - mountPath: /dev
      name: dev-volume
  volumes:
  - name: dev-volume
    hostPath:
      path: /dev
      type: Directory
EOF
```

This Pod will use the following environment variables:

- **DEVICE**: This contains the virtual device where the GPS module is listening. This could be different, depending on the GPS module that you are using. Check the *Further reading* section for more information.

- **CLIENT_ID**: This is the vehicle number that this reader will represent in the system – in this case, 1, the first vehicle.

- **ENDPOINT**: This is the endpoint of the GPS server. You have to use the value obtained in the `GPS_SERVER_IP` variable in the *Deploying gps-server to store GPS coordinates* section.

> **Important Note**
>
> To check the code and create your own container of `gps-reader`, refer to the following link: `https://github.com/sergioarmgpl/containers/tree/main/gps-reader/src`. If you want to use an OLED screen to show information, refer to the following link: `https://github.com/PacktPublishing/Edge-Computing-Systems-with-Kubernetes/blob/main/ch13/code/python/oled.py`. The code uses the OLED included in the Raspberry Pi 4B keyestudio Complete RFID Starter kit.

2. You can check whether your device is reading information by running the following:

```
$ kubectl logs pod/gps-reader -f
```

If you're not sure whether you have access to your device, the way to test is by looking at the frontend and checking whether the device appears on the map.

The output will look like the following:

```
<Response [200]>
{'lat': 11.6318615, 'lng': -80.59205166666666, 'cid': '1'}
```

3. Press *Ctrl* + *C* to cancel.

4. Write `exit` and press *Enter* to exit from your Raspberry.

Now, we have all the backend services running and receiving data, but we need to visualize this information. Let's move to the next section to deploy the `frontend` application.

Visualizing your devices using Open Street Maps in real time

Our application has two parts, one that visualizes the GPS coordinates of the vehicles and their near stops in real time and one that shows the past routes of the vehicle within a time range. So, let's understand first the code of the geo-tracking map showing the devices in real time.

Understanding the geo-tracking map visualizer code

Let's start with a map showing all the vehicles with their coordinates and near stops. We are using HTML, JavaScript, jQuery and the Leaflet library to create the map. Let's look at the code of the map:

```
<!DOCTYPE html>
<html lang="en">
<head>
<Load Javascript libraries>
<Load page styles>
<body>
    <div id='map'></div>
<script>
    <Load Map in an initial GPS position>
    var marker
    var markers = []
    var osm = L.tileLayer(
    'https://{s}.tile.openstreetmap.org/{z}/{x}/{y}.png',
    {
        <Set Open Street Map Initial
        Configuration using Leaflet>
    });

    osm.addTo(map);

    setInterval(() => {
        $.getJSON("URL",
        function(pos) {
            <Delete current markers>
            <Get current positions for each unit or truck>
            <For each position set a marker
             in the map calling
             the function markPosition>
        });
    }, 5000);

    function markPosition(cid,lat,lng,near)
    {
        <Create a maker in the map with
```

```
            Latitude, Longitude, Unit number and near destinies>
      }
</script>
</body>
</html>
```

Our page loads some JavaScript libraries and CSS styles. After that, it loads an initial GPS position to show the map. This map is loaded in the `<div id='map'></div>` code.

The important functions in this code are the following:

- **setInterval**: This function uses jQuery to call the endpoint of `gps-server` to get all the GPS coordinates. To do this, the `setInterval` function calls the `http://GPS_SERVER_IP:3000/clients/positions/unit/km/r/0.1` URL, which returns the current GPS coordinates of each vehicle and their nearby stops in a radius of 0.1 kilometers. To do this, call the `markPosition` function every 5 seconds and send the client ID or vehicle number (`cid`), latency (`lat`), longitude (`lng`), and the `near` variable with the name of the stops. This function creates a mark object in the map.

- **markPosition**: This function creates a Leaflet mark object with a PopUp window in the map. This function also resets the map when it's called.

This application basically loads all the necessary libraries and calls the `setInterval` function to refresh the map every 5 seconds by calling the `markPosition` function. It is important to set an initial GPS position to center the map; this is customizable in the YAML file used to deploy the `frontend` application. Once the map is initialized, it will show all the tracked objects after 5 seconds:

Figure 13.5 – A map showing two tracked devices using GPS

If the connected devices are not sending data to the map, it will show an empty map; in this case, *Figure 13.5* shows two devices connected and sending data. Now, let's say, for example, that we are using the device or vehicle number 2 – in this case, represented as the second CID (client ID). If you click on the blue mark, it will show the current Latitude and Longitude coordinates and the near destinations or stops of the tracked vehicle. In this case, we set two stops, galeno_encinal and la_torre_encinal, which are 0.1 km from the current position of the tracked vehicle. If you click on the blue mark, you will see something like this:

Figure 13.6 – Near destinations showed when clicking the blue mark

This information is calculated every 5 seconds, updating the nearest position of your tracked vehicle in real time. You can customize the code to fit your needs; this is just a quick-start example to build a geo-tracking system using GPS. Let's look at how our vehicles routes report works to show the collected data from the tracked vehicles.

Understanding the vehicles routes report

This application creates a blue line, showing the tracking log stored in MongoDB. This represents the route of the vehicle within a date or time range. Before we take an in-depth look into how it works, let's explore first the code of this page:

```
<!DOCTYPE html>
<html lang="en">
<head>
<Load Javascript libraries>
<Load page styles>
<body>
    <form>
        <input id="cid" name="cid"></input>
        <input id="sdate" name="sdate"></input>
        <input id="edate" name="edate"></input>
        <button onclick="loadMap()"></button>
    </form>
    <div id='map'></div>
<script>
    <Load Map in an initial GPS position>
    var tiles = L.tileLayer(
    'https://{s}.tile.openstreetmap.org/{z}/{x}/{y}.png',
    {
        <Set Open Street Map Initial
        Configuration using Leaflet>
    }).addTo(map);

    function onEachFeature(feature, layer) {
            <Set a popup with the line visualizing the route
            of the vehicle>
    }

    var trip;

    function loadMap(){
        $.getJSON(<DYNAMIC_URL>, function(pos) {
```

```
            var coordinates = [];
            <Creating an array with the coordinates
            between the time range>
            this.trip = {
            <The array with the coordinates and fields
            to visualize in the map>
            };

            var tripLayer = L.geoJSON(this.trip, {
                <Get the trip data and visualize it
                into the map>
            }).addTo(map);
        });
    }
</script>
</body>
</html>
```

Let's analyze the next code sections:

- **trip**: This variable contains all the coordinates to draw a line in the map with the routes covered by the vehicle within a time range.

- **form**: This is an HTML form used to generate the dynamic called to get all the GPS positions between a selected time range.

- **DINAMIC_URL**: This is a dynamic URL used to call `tracking-server` and get all the GPS positions. This URL has the following structure: `http://TRACKING_SERVER_IP:3000/client/2/positions/s/25-05-22-04:39:58/e/25-05-22-04:40:00`.

- **onEachFeature**: This is a function that creates a line with the GPS positions of the vehicle.

- **LoadMap**: This is a function that is called after clicking on the load button of the form to show the routes covered within a time range for a vehicle.

In general, this report page is generated when the **Show Route History** button is clicked, showing the route of the vehicle on the map as follows:

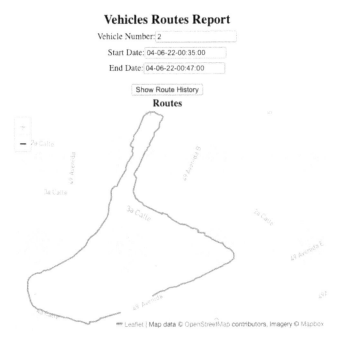

Figure 13.7 – Vehicles Routes Report

Our `tracking-server` service is configured to store and query the log tracking information of the vehicles within a time range using the timestamp captured when data arrives. This application is also configured to use localization times and UTC in different countries. This is a basic implementation of the vehicles routes report that you can customize.

> **Important Note**
> To know more about what is UTC time, you can check the next link: `https://www.timeanddate.com/time/aboututc.html`.

Another feature that this map has is that when you click on the map, it can show some information. In this case, we are showing a sample message, but you can customize it to show additional information, such as the time when the vehicle was in a particular position:

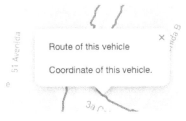

Figure 13.8 – Route information of the vehicle when clicking on the map

To reset the report, you have to reload the page. With this, we are ready to deploy our frontend application, which contains the real time map and this report, so finally, we can access the final application. To do this, let's move on to the next section.

Deploying a real-time map and report application to track your devices

Now we have all the things ready, so we have to deploy the front application that contains our real time map and the report page. To do this, we use a simple Flask application with Python using templates; here is the code:

```
<imported libraries>
<app_initialization>
<CORS configuration>

@app.route("/")
def map():
    return render_template(<Render map.html
                            Using environment variables)
@app.route("/report")
def report():
    return render_template(<Render report.html
                            using environment variables>)


<Starting the application on port 3000>
```

This application renders the map.html page, which loads the Leaflet library to show the maps using the initial latitude and longitude variables. It also sets the endpoint of the gps-server that is called inside this static page. To deploy this application, follow these steps:

1. Create the deployment by running the following:

```
$ cat <<EOF | kubectl apply -f -
apiVersion: apps/v1
kind: Deployment
metadata:
  creationTimestamp: null
  labels:
    app: frontend
```

```
    name: frontend
spec:
  replicas: 1
  selector:
    matchLabels:
      app: frontend
  strategy: {}
  template:
    metadata:
      creationTimestamp: null
      labels:
        app: frontend
    spec:
      containers:
      - image: sergioarmgpl/frontend
        name: tracking-server
        imagePullPolicy: Always
        env:
        - name: LATITUDE
          value: "<YOUR_LATITUDE_COORDINATE>"
        - name: LONGITUDE
          value: "<YOUR_LONGITUDE_COORDINATE>"
        - name: GPS_SERVER
          value: "<YOUR_GPS_SERVER_IP>"
        - name: TRACKING_SERVER
          value: "<YOUR_TRACKING_SERVER_IP>"
        resources: {}
status: {}
EOF
```

This deployment has the following environment variables:

- **LATITUDE**: The initial GPS latitude coordinate to center your map

- **LONGITUDE**: The initial GPS longitude coordinate to center your map

- **GPS_SERVER**: The IP address endpoint of the gps-server application.

- **TRACKING_SERVER**: The IP address endpoint of your tracking-server application.

With these variables, you configure the initial loaded GPS coordinate to center the map and the endpoints to be called by the pages, to show the real-time map of the report routes of the frontend application.

> **Important Note**
>
> To check out the code and create your own container of frontend, refer to the following link: https://github.com/sergioarmgpl/containers/tree/main/frontend/src. To get some initial GPS coordinates to center the map when loading, refer to this website: https://www.gps-coordinates.net.

2. Create a load balancer service for your application:

```
$ cat <<EOF | kubectl apply -f -
apiVersion: v1
kind: Service
metadata:
  creationTimestamp: null
  labels:
    app: frontend
  name: frontend-lb
spec:
  ports:
  - port: 3000
    protocol: TCP
    targetPort: 3000
  selector:
    app: frontend
  type: LoadBalancer
status:
  loadBalancer: {}
EOF
```

3. Get the load balancer IP of our frontend deployment with the following command:

```
$ FRONTEND_IP="$(kubectl get svc frontend-lb
-o=jsonpath='{.status.loadBalancer.ingress[0].ip}')"
```

You can see the value of the FRONTEND_IP environment variable by running the following:

```
$ echo $FRONTEND_IP
```

> **Important Note**
> We used a `LoadBalancer` service type to simplify the implementation, but a cheaper solution would be to use ingress definitions to expose the applications. You can explore the following link for more information: `https://kubernetes.io/docs/concepts/services-networking/ingress`.

4. Now, access your application as `http://<FRONTEND_IP>:3000` in your browser. The important endpoints of the URL are the following:

    - **Geo-tracking map**: `http://<FRONTEND_IP>:3000`

    - **Vehicles routes report**: `http://<FRONTEND_IP>:3000/report`

5. Now, turn on your Raspberry Pi device in your vehicle and wait until your device starts sending GPS coordinates. Don't forget to set your stops for each device. After some seconds or a couple of minutes, your map will start to show your devices in real time.

6. Record some data by driving your equipped vehicle with your Raspberry Pi device and then test your reports.

Now, our simple geolocation system is ready and running. After finishing this chapter, it is important to mention that this is just a basic example that you can extend to fit your needs. Now, it's time to recap what we learned.

Summary

In this chapter, we learned how to take advantage of MongoDB and Redis to store and query GPS coordinates to build a basic geolocalization system. We also learned how to integrate a GPS module to an edge device and send information to the cloud to finally visualize how a vehicle is moving in real time on a map, showing the near stops in a circle area, and simulating in that way a basic tracking delivery system. This shows how to implement a simple use case using geolocalization and how edge devices moving in real time interact in a geolocalization system. In the next chapter, we are going to learn how to use machine learning and computer vision to create a small smart traffic project.

Questions

Here are a few questions to validate your new knowledge:

- How can I use GPS technologies to create a geolocalization system?

- How can I use Redis to store GPS coordinates and do queries with this data?

- How can I use MongoDB to store logs for a geolocalization system?

- How can I design a real-time application that shows the GPS positions of moving vehicles?

- How can I use edge computing and K3s to create a distributed system to track vehicles?

Further reading

You can refer to the following references for more information on the topics covered in this chapter:

- VK-162 G-Mouse GPS module: https://www.amazon.com/Navigation-External-Receiver-Raspberry-Geekstory/dp/B078Y52FGQ

- Redis geospatial index commands: https://redis.io/commands/?group=geo

- Geospatial data: https://www.mongodb.com/docs/manual/geospatial-queries

- *Positioning chips and modules*: https://www.u-blox.com/en/positioning-chips-and-modules

- *Basics of Hash Tables*: https://www.hackerearth.com/practice/data-structures/hash-tables/basics-of-hash-tables/tutorial

- *Find Arduino Port on Windows, Mac, and Linux*: https://www.mathworks.com/help/supportpkg/arduinoio/ug/find-arduino-port-on-windows-mac-and-linux.html

- raspi-config: https://geek-university.com/raspi-config

- *GPS – NMEA sentence information*: http://aprs.gids.nl/nmea

- *Leaflet – an open source JavaScript library for mobile-friendly interactive maps*: https://leafletjs.com

- *GPS Coordinates*: https://www.gps-coordinates.net

- *Epoch and Unix Timestamp Conversion Tools*: https://www.epochconverter.com

- pytz timezones Library: https://pypi.org/project/pytz

- Country codes: https://www.iban.com/country-codes

14

Computer Vision with Python and K3s Clusters

Artificial intelligence (**AI**) is commonly used to substitute activities that humans do every day. It can give systems the intelligence to operate autonomously without human intervention in most cases. **Computer vision** (**CV**) is a subcategory of AI that focuses on detecting objects in videos and images. CV is often used to detect traffic in a city. This chapter focuses on building a basic smart traffic system that consists of detecting objects such as cars, trucks, and pedestrians when a vehicle is moving. For this, the system uses the OpenCV, TensorFlow, and scikit-learn Python libraries and a camera to perform computer vision at the edge on a Raspberry Pi. This system also shows locally to drivers a map within the detected objects, and it also implements a public map for global detected object visualization. This public map can be used as a real-time traffic state map that municipalities can use.

In this chapter, we're going to cover the following main topics:

- Computer vision and smart traffic systems
- Using Redis to store temporary object **Global Positioning System** (**GPS**) positions
- Deploying a computer vision service to detect car obstacles using OpenCV, TensorFlow Lite, and scikit-learn
- Deploying the edge application to visualize warnings based on computer vision
- Deploying a global visualizer for the smart traffic system

Technical requirements

To deploy our computer vision system in this chapter, you will need the following:

- A Kubernetes cluster hosted in your public cloud provider (**Amazon Web Services** (**AWS**), Azure, **Google Cloud Platform** (**GCP**)).

- A Raspberry Pi 4B with an 8-GB micro **Secure Digital** (**SD**) card with a small-monitor **liquid-crystal display** (**LCD**) screen to use in a car.

- A Logitech C922 PRO webcam, recommended because of its quality and support on Linux.

- Multiple VK-162 G-Mouse USB GPS Dongle Navigation modules, for your edge Raspberry devices.

- Basic knowledge of AI.

- `kubectl` configured to be used in your local machine for your Kubernetes cloud cluster to avoid using the `--kubeconfig` parameter.

- Clone the `https://github.com/PacktPublishing/Edge-Computing-Systems-with-Kubernetes/tree/main/ch14` repository if you want to run the **YAML Ain't Markup Language** (**YAML**) configuration by using `kubectl apply` instead of copying the code from the book. Take a look at the `python` directory inside the `code` directory and the `yaml` directory for YAML configurations that are inside the `ch14` directory.

With this, you can deploy Prometheus and Grafana to start experiment monitoring in edge environments.

Computer vision and smart traffic systems

AI is an area of computer science that consists of simulating human intelligence using mathematics, statistics, linguistics, computer science, and other sciences. AI can also be defined as the study of rational agents, as depicted in the following diagram:

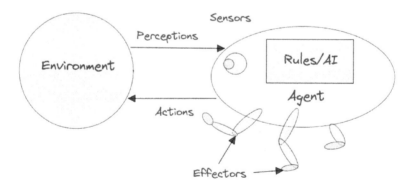

Figure 14.1 – Agents

Taking *Figure 14.1* as a reference, an agent receives perceptions coming from the environment. These perceptions are captured by sensors, and this information is processed to perform an action using effectors. Actions are decided by internal rules installed inside the agent. These actions involve the use of effectors such as arms, legs, or wheels, for example.

These internal rules can be implemented using different **machine learning (ML)** paradigms such as **supervised learning (SL)**, **unsupervised learning (UL)**, and **reinforcement learning (RL)**.

ML is a type of AI that uses historical data as input to do predictions. Computer vision is a subset of ML applied to image and video analysis using predictions. In our chapter, we are going to do predictions about what our agent is capturing using a camera and take decisions according to that information, but we are going to apply computer vision to create a smart traffic system. Let's have a look at the following diagram, which shows how our system will be implemented to create a smart traffic system using computer vision at the edge:

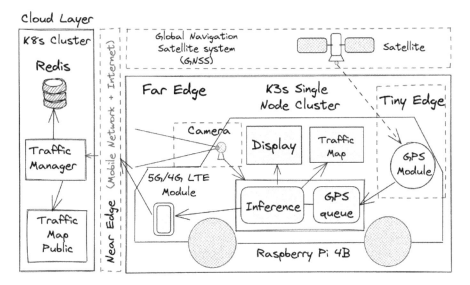

Figure 14.2 – Smart traffic system using computer vision

Smart traffic systems are often used by municipalities to improve safety, and traffic flow on streets in a cost-effective way. Our system can be used in two modes. The static mode uses a camera in a static location point in the city, and the dynamic mode uses a car to scan traffic where the car is moving. We are going to use the dynamic mode. Now, let's explain our system using the layers of the edge computing systems, as follows:

- **Cloud layer**: Here, we are going to use an **application programming interface (API)** called **Traffic Manager** that stores all detected objects at the edge in a Redis instance. The data stored will contain the type of object—car, truck, and person—which represents a level 1 warning on our system and the GPS coordinates. This means that a vehicle driver will be warned of previously detected objects by other drivers. Our API will store the GPS position of these objects, which potentially could be obstacles for a vehicle. This layer will also include a frontend application called **Traffic Map Public** that shows the objects detected on a map. This application could be used by the municipality to monitor all traffic across the city.

- **Near edge**: This layer has the **fourth-generation (4G)/fifth-generation (5G) Long-Term Evolution (LTE)** mobile network used to send information to the internet. This layer will transport information collected at the edge to send it to the cloud layer.

- **Far edge**: Our far edge has a Raspberry Pi that will process the information captured by a camera. This device has installed K3s as a single node cluster to manage all services that the system uses. K3s can brings automation to the system. K3s can easily update and maintain the system and can extend the system to use more nodes. These additional nodes can be used to add multiple cameras for object detection at multiple angles. The computer vision application that runs in the cluster consists of two displays and two APIs. One display runs outside K3s but in the same device as a Python script, and it's the service that captures the video. This service consists of a Python program that captures video and detects objects using OpenCV and a precompiled model for TensorFlow Lite for object detection. Here is where computer vision occurs. The system uses a small LCD touchscreen connected to the device. The other display is a frontend application that runs on a browser; it shows detected objects across a map, not only showing these locally but also showing all detected objects by all vehicles in a radius of 500 meters. Detected objects will be classified by the Inference API, which classifies objects according to their level of warning for a driver. These warnings are represented at three levels: levels 1 and 2 represent a warning, and level 3 could be ignored as an obstacle for a driver. The Inference API contains a precompiled decision tree to do classification. The **GPS Queue** API manages all GPS coordinates and periodically sends information about detected objects that represent a warning to the cloud to be shown to other drivers. The whole application uses the Display, Traffic Map, Inference, and GPS Queue components to process and visualize detected objects. The GPS Queue service is based on the GPS service created in *Chapter 5, K3s Homelab for Edge Computing Experiments*, with some modifications. Something important to consider is that you can accelerate your object detection by using an external device that accelerates **neural network (NN)** processing. Some devices that you can consider are the Coral USB Accelerator from Google, the Rock Pi neural compute stick **Universal Serial Bus (USB)**, and the NVIDIA Jetson Nano. These devices accelerate the NN processing of OpenCV by delegating processing to a dedicated processing unit sometimes called a **graphics processing unit (GPU)** or a **Tensor Processing Unit (TPU)**. The OpenCV library uses TensorFlow Lite models, so the use of these devices can increase the number of **frames per second (FPS)** analyzed that have some GPU that can be used by TensorFlow Lite, which is designed to run on edge devices to accelerate your video analysis. For more information, check the *Further reading* section.

- **Tiny edge**: Here, we can find an LCD screen to display all detected objects in real time and warnings for the driver. You can also find the VK-162 G-Mouse GPS module here.

To summarize this workflow, our vehicle first captures images with its camera; then, the video frames or images are captured using OpenCV and classified using TensorFlow Lite, then are classified according to their level of warning representation for the drivers by the Inference API. This information is shown locally in the LCD and browser. The GPS coordinate data sent to the cloud is shown in a public web frontend application in the cloud. So now, let's get started in building a basic smart traffic system to alert drivers.

Using Redis to store temporary object GPS positions

We are going to use Redis to store our GPS coordinates for all detected objects using computer vision. This is a basic configuration to deploy Redis for this purpose. This Redis instance must be deployed in the cloud. As we explained in *Chapter 13, Geolocalization Applications Using GPS, NoSQL, and K3s Clusters*, we are going to use a geospatial index to represent our data. The difference will be that we are going to implement temporary storage of data using a **time-to-live** (TTL) feature that auto-expires keys in Redis. For this, we are going to continuously watch hash keys in Redis if they still exist. For each detected object, the type and level of warning are stored in a hash key, and a coordinate will be added in a geospatial sorted set. Then, a TTL is configured for the hash key. If this hash key expires, it will be removed from a geospatial set called `traffic`, which stores all traffic objects detected by other drivers. In this way, we implemented a kind of garbage functionality to remove old detected objects during traffic hours. The reason is that the detected objects are relevant just for a certain amount of time, then have to be deleted. So, let's install our Redis deployment by following the next steps:

1. Create a **PersistentVolumeClaim** for Redis to persist our data, like so:

    ```
    $ cat <<EOF | kubectl apply -f -
    apiVersion: v1
    kind: PersistentVolumeClaim
    metadata:
      name: db-pv-claim
    spec:
      accessModes:
        - ReadWriteOnce
      resources:
        requests:
          storage: 5Gi
    EOF
    ```

2. Now, create a **ConfigMap** to configure Redis to use an authentication password, as follows:

    ```
    $ cat <<EOF | kubectl apply -f -
    apiVersion: v1
    kind: ConfigMap
    metadata:
      name: redis-configmap
    ```

```
data:
  redis-config: |
    dir /data
    requirepass YOUR_PASSWORD
EOF
```

3. Create a deployment for Redis using the previous `redis-configmap` **ConfigMap** and the `db-pv-claim-1` **PersistentVolumeClaim** with some resource limits, using the following command:

```
$ cat <<EOF | kubectl apply -f -
apiVersion: apps/v1
kind: Deployment
metadata:
  creationTimestamp: null
  labels:
    app: redis
  name: redis
spec:
  replicas: 1
  selector:
    matchLabels:
      app: redis
  strategy: {}
  template:
    metadata:
      creationTimestamp: null
      labels:
        app: redis
    spec:
      containers:
      - name: redis
        image: redis:6.2
        command:
          - redis-server
          - /redisconf/redis.conf
        ports:
```

```
            - containerPort: 6379
          resources:
            limits:
              cpu: "0.2"
              memory: "128Mi"
          volumeMounts:
          - mountPath: "/data"
            name: redis-storage
          - mountPath: /redisconf
            name: config
        volumes:
        - name: config
          configMap:
            name: redis-configmap
            items:
            - key: redis-config
              path: redis.conf
        - name: redis-storage
          persistentVolumeClaim:
            claimName: db-pv-claim-1
  status: {}
EOF
```

4. Now, create a service for Redis opening port 6379, like so:

```
$ cat <<EOF | kubectl apply -f -
apiVersion: v1
kind: Service
metadata:
  labels:
    app: redis
  name: redis
spec:
  ports:
  - port: 6379
    protocol: TCP
    targetPort: 6379
```

```
    selector:
      app: redis
    type: ClusterIP
  EOF
```

We now have Redis installed. Let's move on to deploying our computer vision service at the far edge, in the next section.

Deploying a computer vision service to detect car obstacles using OpenCV, TensorFlow Lite, and scikit-learn

In this section, we are going to explore how to configure the object detection system that runs at the edge with all its components. This section also shows how to configure the public web application running in the cloud that stores and shows information about all detected objects at the edge. Let's start by first configuring our Raspberry Pi device in the next section.

Preparing your Raspberry Pi to run the computer vision application

Before installing our software, we have to prepare our device to run it. For this, let's start to configure our Raspberry Pi 4B following the next steps:

1. Install Raspbian Pi OS (32 bit) using Debian Bullseye, released at least from 2022-04-04. The code to run the TensorFlow Lite model in this chapter has to run on an ARMv7 device to support the Coral USB Accelerator device and the LCD screen. ARM64 is not supported yet.

2. Depending on your webcam, you have to install drivers. In this case, we are using the Logitech C922 PRO webcam, which is automatically detected by Raspbian.

3. Connect and configure your GPS module. In this case, our VK-162 G-Mouse module is autodetected by Raspbian too.

4. Configure the network to use a wireless connection, to install all the necessary packages to run the application. Later, you can reconfigure your wireless connection to connect to your access point in your smartphone, but you have to delete the previous connection in the /etc/wpa_supplicant/wpa_supplicant.conf file.

5. Install the drivers of your LCD screen. In this case, we are using the Miuzei **High-Definition Multimedia Interface** (**HDMI**). This will flip the screen horizontally and activate the touch feature (this will be the last step once all the things are configured). You can check the repository at https://github.com/goodtft/LCD-show.git, and you can use any LCD screen.

6. Before installing K3s, remember to activate the CGROUPS in the /boot/cmdline.txt file, then add the next flags at the end of the line:

    ```
    cgroup_memory=1 cgroup_enable=memory
    ```

> **Important Note**
> For more information about CGROUPS visit this link: `https://man7.org/linux/man-pages/man7/cgroups.7.html`

7. Get your current **Internet Protocol** (**IP**) address by running `ifconfig`, then take a look at the `wlan0` interface, as follows:

    ```
    $ ifconfig wlan0
    ```

8. Install K3s by running the following command:

    ```
    $ MASTER_IP=YOUR_PRIVATE_IP
    $ curl -sfL https://get.k3s.io | INSTALL_K3S_EXEC="--
    write-kubeconfig-mode 644" sh -s -
    ```

9. Now, you can test if everything is working by running the following command:

    ```
    $ kubectl get nodes
    ```

    This will return your unique node running.

Now, our edge device is ready to be used to run our service that performs computer vision at the edge. For this, let's move on to the next section.

Deploying the inference service to detect objects

The `inference` service is used in this scenario to do predictions and to classify if an object represents an obstacle for a driver. We use the next table for that:

object	N	warning_level
car	1	1
cat	2	2
person	3	1
dog	4	2
semaphore	5	1
truck	6	1
other	1000	3

For example, a car identified by the id 1 in the n field represents a level 1 of warning, so all the objects with `warning_level` equal to 1 or 2 will be recorded as potential objects that can obstruct traffic or represent danger for the driver. If an object is classified with the value 1000, the object doesn't represent any danger, so it is not recorded.

The source code of this service consists of two files: index.py and create_model.py. The index.py file contains a basic API to return predictions by calling the model to predict using the /predict path. It has basic code to load the precompiled ML model. The create_model.py file contains code to train and generate a model that will be used for this API using index.py. The code looks like this:

```
import pandas as pd
from sklearn import tree
from joblib import dump


df = pd.read_csv("safety_rules.csv",sep=',', header='infer',
encoding='latin-1')
df = df.drop(['object'], axis=1)
df.head()


feature_cols = ["n"]
X = df.loc[:, feature_cols]
y = df.warning_level


clf = tree.DecisionTreeRegressor()
model = clf.fit(X, y)


dump(clf, 'safety_rules.model')
```

Here, we read our safety_rules.csv **comma-separated values** (CSV) file with the rules inside. After, that the information in this file is converted into a DataFrame and the column object is removed from this DataFrame using drop. In AI, you have to represent texts as values. Our object column has a numeric representation in the n column, so the column object can be ignored. The data loaded from the CSV file is represented as a Pandas DataFrame that is used in scikit-learn as the source of data to generate a decision tree. A decision tree is an ML algorithm that can use classified data to do predictions using the data structure of trees for predictions. So, it is one of the simplest methods to do predictions using ML. After the DataFrame is loaded, scikit-learn does its training processes to generate a safety_rules.model model that could be used later in the API for predictions. Every time you build the container, the model is updated by calling the create_model.py file inside the Dockerfile of this API. Now, the serving code for the API will look like this:

```
<Import Flask and Scikit Learn libraries>


def loadModel():
    <Load the model safety_rules.model>
```

```
<Assign the loaded model to the variable clf>

@app.route('/predict', methods=["POST"])
def predict():
    <Use clf variable to call the prediction method>
    <Return the prediction using JSON format>


<Inference service initialization on port 3000 by default>
```

By calling the /predict **Uniform Resource Locator** (URL), you can get predictions from the model based on the rules set in the safety_rules.csv file. You can add more values to classify your images by adding new values in the file and regenerating the container with the new model.

> **Important Note**
> To check the code and update the model, check the next link: https://github.com/sergioarmgpl/containers/tree/main/inference/src.

Now, let's deploy our inference service in our **Advanced RISC Machine** (ARM) device by following the next steps:

1. Create a deployment for the inference API, as follows:

    ```
    $ cat <<EOF | kubectl apply -f -
    apiVersion: apps/v1
    kind: Deployment
    metadata:
      creationTimestamp: null
      labels:
        app: inference
      name: inference
    spec:
      replicas: 1
      selector:
        matchLabels:
          app: inference
      strategy: {}
      template:
        metadata:
    ```

```
              creationTimestamp: null
              labels:
                app: inference
          spec:
            containers:
            - image: sergioarmgpl/inference
              name: inference
              imagePullPolicy: Always
              resources: {}
      status: {}
      EOF
```

2. Let's port forward the service running, like so:

```
$ kubectl port-forward --address 0.0.0.0 deploy/inference
3000:3000
```

3. Now, let's call the `inference` API to get some predictions. Let's use an object detected and classified as `other` with the number 6; it will return a warning level of 3 based on the prediction table. The code is illustrated in the following snippet:

```
$ curl --header "Content-Type: application/json" \
--request POST --data '{"data":[6]}' \
http://localhost:3000/predict
```

This will return the following output:

```
{
   "prediction": 3.0
}
```

Our inference service is now running, ready to be called inside our device to classify the detected images. Let's continue deploying the gps-queue service in the next section.

Deploying the gps-queue service to store GPS coordinates

The gps-queue service is composed of several containers dedicated to a specific task. First, initialize an init container called init-gps-queue that adds an initial value of -1 inside the /tmp/gps file. This file stores the last GPS coordinate generated. Then, the gps-queue container is in charge of reading the GPS coordinates from our GPS module, so it needs permission to access the /dev folder from the host. Once the GPS coordinate is read, it is stored in /tmp/gps. After this, the sync-traffic-events container calls the gps-api container every 30 seconds by default using the `http://localhost:3000/traffic` endpoint, which sends the detected objects with their warning classification and GPS coordinate to the `http://<TRAFFIC_MANAGER_IP>:5000` public endpoint, which stores this information for some time to be shown in the `traffic-map-public` service that has public access to show the objects detected by other vehicles. Before deploying our service, let's explore a little bit the code of the `gps-queue` container, as follows:

```
<Import necessary Python libraries to read the GPS module>

<cid variable to set a unique client id for these coordinates>
<device variable to set where the GPS module will be read in /
dev>
<ser variable to configure the serial communication with the
GPS module>

<Initializing the device to read information>

while True:
    <Read the Coordinate and store it into /tmp/gps>
```

This code configures the GPS module and stores the coordinate in the `/tmp/gps` file, which is shared by the `gps-queue` and `gps-api` containers. It uses a `cid` variable to associate each GPS coordinate with a unique client **identifier (ID)** that could be used for customizations to create your own system. The information will be stored in the next format:

```
{'lat': <LATITUDE_VALUE>,'lng':<LONGITUDE_VALUE>,'cid':<CLIENT_
ID>}
```

Now, let's explore the code inside the `gps-api` container, as follows:

```
<Import the necessary Python libraries to run this code>
<Set traffic_events variable to accumulate detected objects for
a time period>
<Flask and CORS configuration>

@app.route("/gps", methods=["GET"])
def getGPSCoordinate():
    <Read coordinate form /tmp/gps>
    <Return the GPS coordinate as JSON as
    {'lat': <LATITUDE_VALUE>,'lng':<LONGITUDE_VALUE>
    ,'cid':<CLIENT_ID>}
```

```
>

@app.route("/traffic/event", methods=["POST"])
def registerTrafficEvent():
    <Read last GPS coordinate from /tmp/gps>
    <Get object type and warning classification
     from the computer vision service>
    <Generate the Timestamp value for the new detected object>
    <Assign to a variable the warning, Latitude, Longitude
     and timestamp information for the object>
    <Add this information to the traffic_events array
     to store it temporary the value>
    <Return the object ide and that the request was processed>


@app.route("/traffic", methods=["GET"])
def syncTrafficEvents():
    <Filter similar objects stored in the
     traffic_events array>
    <Send the filtered array using JSON format to the
     endpoint http://<TRAFFIC_MANAGER:5000>/traffic/1
     to store this information and get it locally and
     public by calling the endpoint
     http://<TRAFFIC_MANAGER:5000>/traffic>
    <Return that the information syncTrafficEvents
     was processed>


<GPS Queue service initialization on port 3000 by default>
```

As an explanation, the /gps path of this API returns the value of the last GPS coordinate stored in /tmp/gps, and the /traffic/event path receives the object detected from the edge device running the detect.py program. This happens every second. Then, the information is stored temporarily in the traffic_events array. Inside the Pod, the sync-traffic-events container calls the /traffic endpoint of the API running inside the gps-api container, which filters the traffic_events array to have just unique objects detected because the edge program gets a maximum of eight detected objects per video-frame analysis. Once the array is filtered, it is sent to the **Traffic Manager** service that is running in the cloud by calling its endpoint at http://<TRAFFIC_MANAGER:5000>/traffic/1. This information is requested later by the **Traffic Map Public** web application using the http://<TRAFFIC_MANAGER:5000>/traffic URL, which shows the globally stored objects detected from all the devices in a map using the Leaflet library.

To deploy this service, execute the following steps:

1. Create a deployment for the GPS queue, like so:

```
$ cat <<EOF | kubectl apply -f -
apiVersion: apps/v1
kind: Deployment
metadata:
  labels:
    app: gps-queue
  name: gps-queue
spec:
  replicas: 1
  selector:
    matchLabels:
      app: gps-queue
  template:
    metadata:
      labels:
        app: gps-queue
    spec:
      initContainers:
      - image: busybox:1.34
        name: init-gps-queue
        command: ['sh', '-c', "echo '-1' >> /tmp/gps"]
        securityContext:
          runAsUser: 1
        volumeMounts:
        - name: tmp
          mountPath: /tmp
      containers:
      - image: sergioarmgp1/gps_queue
        name: gps-queue
        imagePullPolicy: Always
        env:
        - name: DEVICE
```

```yaml
        value: "/dev/ttyACM0"
      securityContext:
        privileged: true
        capabilities:
          add: ["SYS_ADMIN"]
      volumeMounts:
      - mountPath: /dev
        name: dev-volume
      - name: tmp
        mountPath: /tmp
    - image: sergioarmgpl/gps_api
      name: gps-api
      ports:
      - containerPort: 3000
      imagePullPolicy: Always
      env:
      - name: ENDPOINT
        value: "http://<TRAFFIC_MANAGER_IP>:5000"
      securityContext:
        runAsUser: 1
      volumeMounts:
      - name: tmp
        mountPath: /tmp
    - image: curlimages/curl
      name: sync-traffic-events
      env:
      - name: URL
        value: "http://localhost:3000/traffic"
      - name: DELAY
        value: "30"
      command: [ "sh", "-c"]
      args:
      - while :; do
          curl ${URL};
          sleep ${DELAY};
```

```
        done;
    volumes:
    - name: dev-volume
      hostPath:
        path: /dev
        type: Directory
    - name: tmp
      emptyDir: {}
status: {}
```

> **Important Note**
>
> To check the code and create your own containers, you can check the next links:
>
> `https://github.com/sergioarmgpl/containers/tree/main/gps-api/src` and `https://github.com/sergioarmgpl/containers/tree/main/gps-queue/src`

Let's pay attention to the variables that this deployment uses in its containers. These are explained in more detail here:

- `gps-queue`:

  - `DEVICE`: Configures the device where your GPS module is detected. For the VK-162 G-Mouse module, the default value used is `/dev/ttyACM0`.

- `gps-api`:

  - `ENDPOINT`: Configures the public endpoint where all detected objects with GPS coordinates and warnings are stored. This is the public service that stores the coordinates. By default, this is `http://<TRAFFIC_MANAGER_IP>:5000`.

- `sync-traffic-events`:

  - `URL`: Contains the local URL called periodically to send information about all detected objects. This will call the API configured in the `gps-api` container. By default, this is `http://localhost:3000/traffic`.

  - `DELAY`: Configures the amount of time to wait to send the last objects detected with their information. By default, this is 30, which represents the time in seconds.

These values could be used to customize the behavior of the service that processes the objects detected and its GPS coordinates.

2.   If you want to test the endpoints of this service, you can run inside your edge device `port-forward` to access the API using the `curl` command, like so:

```
$ kubectl port-forward --address 0.0.0.0 deploy/gps-queue
3001:3000
```

For example, you can execute the following command:

```
$ curl http://localhost:3001/gps
```

It will return something like this:

```
{'lat': <LATITUDE_VALUE>,'lng':<LONGITUDE_VALUE>
  ,'cid':<CLIENT_ID>}
```

We have now deployed the `gps-queue` service and it's ready to be used. It's time to deploy our local web application that will show detected objects at the edge using our edge device equipped with a camera. For this, we have to solve the **Cross-Origin Resource Sharing** (**CORS**) restriction call that happens when it calls the `traffic-manager` public API from the local `traffic-map` application. CORS is a mechanism that allows or restricts resources on a web page to be requested from a domain outside the current one. In this scenario, it's called a public API from a local web application. So, let's move on to the next section to create a simple proxy to resolve this issue.

Deploying traffic-manager to store GPS coordinates

The `traffic-manager` service receives detected objects with their GPS coordinates and warning-level classification. This API runs in the cloud, and it's called periodically by the edge device while it's moving and detecting objects. This service consists of two containers: one that gives an API to recollect objects detected, and another that is in charge of auto-expiring detected objects and global traffic information. This is because traffic is constantly changing during the day. You can configure these values to fit your own scenario. Let's explore first the code of the API in the `traffic-manager` container, as follows:

```
<Import the necessary Python libraries to run this code>

<Flask and CORS configuration>

<Set time to expire the traffic and objects by setting the
values of the variables ttl_trf, ttl_obj>

def redisCon():
    <Set and return the Redis connection>

@app.route("/traffic/1", methods=["POST"])
```

```
def setBulkTrafficObjects():
   <Get the Redis connection calling redisCon()>
   <Get detected objects from the POST request>
   <Omit to store similar detected objects in a
   5 meters radius>
   <Set a hash value to store type and warning
    level for each object>
   <Set expiring time for each hash stored>
   <Return that the operation was successful
{"setTrafficObject":"done"}>


@app.route("/traffic/unit/<unit>/r/<radius>"+
"/lat/<lat>/lng/<lng>", methods=["GET"])
def getTrafficObjects(unit,radius,lat,lng):
   <Get the Redis connection calling redisCon()>
   <Get the objects detected and its metadata
   from the previous stored hash
   in the radius configured in the request>
   <Return that the operation was successful and
   the objects found
   in the next format:
   {"getTrafficObjects":"done",
    "objects":data
   }>


<Service initialization on port 3000 by default>
```

This container has two endpoints with the /traffic/1 path. This service stores detected objects at the edge by creating a hash key with the form object:<object-id>:data that stores the type and the warning level, and in the traffic geospatial set stores the GPS coordinate. An expiration time to the traffic key is set or renewed, and for the new object:<object-id>:data hash key, the expiration time is set too. After calling the /traffic/unit/<unit>/r/<radius>/lat/<lat>/lng/<lng> path, the call returns near detected objects in the radius defined in the request. This is a public service that all the edge devices will access periodically to send updates of objects detected while they are moving. Now, let's explore the code of the autoexpire container, as follows:

```
<Import all the necessary libraries>
<Set Redis connection in an r variable>
```

```
while True:
    <Get all the objects inside the traffic sorted set>
    <Check if each member of the set has its hash value>
    <If not remove the member of the sorted set>
    <Wait until the configured delay ends to
    Update the set again>
```

This container basically checks if each member of the traffic geospatial set has metadata available in the `object:<object-id>:data` hash key. If none exists, this means that the object passed the maximum amount of time to be relevant in the traffic, which means that it has expired too, and then this code removes the member from the sorted set. This process is called periodically after waiting for a certain number of seconds that are configured by the DELAY variable.

To deploy the `traffic-manager` service, proceed as follows:

1. Create a deployment for the GPS server, like so:

    ```
    $ cat <<EOF | kubectl apply -f -
    apiVersion: apps/v1
    kind: Deployment
    metadata:
      creationTimestamp: null
      labels:
        app: traffic-manager
      name: traffic-manager
    spec:
      replicas: 1
      selector:
        matchLabels:
          app: traffic-manager
      strategy: {}
      template:
        metadata:
          creationTimestamp: null
          labels:
            app: traffic-manager
        spec:
          containers:
          - image: sergioarmgpl/autoexpire
    ```

```
            name: autoexpire
            imagePullPolicy: Always
            env:
            - name: REDIS_HOST
              value: "redis"
            - name: REDIS_AUTH
              value: "YOUR_PASSWORD"
            - name: DELAY
              value: "30"
          - image: sergioarmgpl/traffic_manager
            name: traffic-manager
            imagePullPolicy: Always
            env:
            - name: REDIS_HOST
              value: "redis"
            - name: REDIS_AUTH
              value: "YOUR_PASSWORD"
            - name: TTL_TRAFFIC
              value: "900"
            - name: TTL_OBJECT
              value: "180"
            resources: {}
      status: {}
      EOF
```

This deployment uses the following variables:

- REDIS_HOST: This is the name of the Redis service. This variable can be customized to fit your needs.

- REDIS_AUTH: This is the password to connect to the Redis service.

- TTL_TRAFFIC: This is the URL of the tracking-server service. In this case, the URL matches the internal tracking-server service on port 3000.

- TTL_OBJECT: This is the URL of the tracking-server service. in this case, the URL matches the internal tracking-server service on port 3000.

- DELAY: This is the time to wait to check if a member inside the traffic geospatial sorted set expired.

By configuring these variables, you can customize the behavior of this deployment.

2. Now, let's create a service for this deployment as a **LoadBalancer**. This IP address will be used in our edge device to propagate this information in the cloud to be accessible to all drivers that use this smart traffic system. The code is illustrated in the following snippet:

```
$ cat <<EOF | kubectl apply -f -
apiVersion: v1
kind: Service
metadata:
  labels:
    app: traffic-manager
  name: traffic-manager-lb
spec:
  ports:
  - port: 5000
    protocol: TCP
    targetPort: 3000
  selector:
    app: traffic-manager
  type: LoadBalancer
EOF
```

3. Get the load balancer IP address for your `traffic-manager` deployment with the following command:

```
$ TRAFFIC_MANAGER_IP="$(kubectl get svc traffic-manager-lb  -o=jsonpath='{.status.loadBalancer.ingress[0].ip}')"
```

You can see the value of the TRAFFIC_MANAGER_IP environment variable by running the following command:

```
$ echo $TRAFFIC_MANAGER_IP
```

Note that it takes some time after the IP address of the load balancer is provisioned. You can check the state of the services by running the following command:

```
$ kubectl get svc traffic-manager-lb
```

Wait until the EXTERNAL_IP environment variable is provisioned.

Also, take note that the $TRAFFIC_MANAGER_IP value will be used to configure the proxy service in the edge device.

4. (*Optional*) If you want to test this API to insert an object manually, run the following command:

```
$ curl -X POST -H "Accept: application/json" \
-H "Content-Type: application/json" \
--data '{
    "object":"person",
    "warning":1,
    "position":{"lat":1.633518,"lng": -90.591706}
}' http://$TRAFFIC_MANAGER_IP:3000/traffic/1
```

This will return the following output:

```
{
    "setTrafficObject": "done"
}
```

5. (*Optional*) To get all detected objects in a radius of 0.1 kilometers, run the following command:

```
$ curl -X GET -H "Accept: application/json" \
http://$TRAFFIC_MANAGER_IP:3000/traffic/objects/unit/
km/r/0.1/lat/1.633518/lng/-90.5917
```

This will return the following output:

```
{
    "getTrafficObjects": [
        "person"
    ]
}
```

Now, our traffic-manager API is running in the cloud. Let's move on to use this API in our edge device using a proxy to prevent CORS restrictions when calling the API, in the next section.

Deploying a simple proxy to bypass CORS

The proxy service is used to bypass the CORS restriction that occurs when a local website running on a private network tries to call a public API using a public API address. Using a proxy to forward requests to this public site could be one possible and simple solution to solve this. Another one is to modify the request headers on the API call and add the necessary headers to bypass the CORS restriction. In this case, we are going to use a proxy build with Flask to forward all local GET requests to the `traffic-manager` API, which is a public API deployed in the cloud and is accessible over the internet. Let's explore the code a little bit before deploying the `proxy` service, as follows:

```python
from flask import Flask,request,redirect,Response
import os
import requests
app = Flask(__name__)
url = os.environ['URL']

@app.route('/<path:path>',methods=['GET'])
def proxy(path):
    global url
    r = requests.get(f'{url}/{path}')
    excluded_headers = ['content-encoding'
    , 'content-length', 'transfer-encoding'
    , 'connection']
    headers = [(name, value) for (name, value) in
    r.raw.headers.items() if name.lower() not in
    excluded_headers]
    response = Response(r.content, r.status_code, headers)
    return response

if __name__ == '__main__':
    app.run(debug = False,port=5000)
```

This code basically receives all GET requests on any path and forwards the requests with all the important headers to the URL defined in the environment variable. This API is accessible using port 5000. Now, let's move on to deploy this simple proxy to forward all calls from our local **Traffic Map** web application to the public **Traffic Manager** service as though it is running locally in the same host where **Traffic Map** is running. To deploy the `proxy` service, execute the following steps:

1. Create a deployment for the GPS server, like so:

    ```
    $ cat <<EOF | kubectl apply -f -
    apiVersion: apps/v1
    kind: Deployment
    metadata:
      creationTimestamp: null
    ```

```
    labels:
      app: proxy
    name: proxy
  spec:
    replicas: 1
    selector:
      matchLabels:
        app: proxy
    strategy: {}
    template:
      metadata:
        creationTimestamp: null
        labels:
          app: proxy
      spec:
        containers:
        - image: sergioarmgpl/proxy
          name: proxy
          imagePullPolicy: Always
          env:
          - name: URL
            value: "http://<TRAFFIC_MANAGER_IP>:5000"
          resources: {}
  status: {}
  EOF
```

This deployment uses the following variables:

- URL: This variable has the URL where the proxy is going to redirect all GET requests received by the proxy in port 5000. This URL will be the traffic-manager public IP address using the format http://<TRAFFIC_MANAGER_IP>:5000.

> **Important Note**
> To check the code and create your own container, you can check the next link: https://github.com/sergioarmgpl/containers/tree/main/proxy/src. This small proxy is a custom implementation that you can implement using languages other than Python to have all the control in your implementation. You can also use solutions such as using NGINX with a proxy_pass configuration, and so on.

2. You can test the proxy by running something like this:

    ```
    $ curl http://localhost:5000/<REMOTE_PATH>
    ```

    Here, the remote path could be /traffic, which is a URL where the **Traffic Manager** service returns all objects globally detected by drivers.

Now our proxy is running, let's deploy our **Traffic Map** web application to show the detected objects that represent warnings for drivers in the next section.

Deploying the edge application to visualize warnings based on computer vision

Our visual application consists of two parts: the first one is a web application that shows all data from all drivers using the smart traffic system, and the other one is a desktop application that shows the detected objects in real time. So, let's start installing our web application to visualize objects detected by different drivers in the next section.

Installing the Traffic Map application to visualize objects detected by drivers

We have now set up the necessary APIs to visualize what our device detected. We have to continue deploying our web application to visualize this object on a map. This is where our **Traffic Map** application comes in handy. But let's explore the code first before deploying it, as follows:

```
<imported libraries>
<app_initialization>
<CORS configuration>
@app.route("/")
def map():
    return render_template(<Render map.html
                            Using environment variables
                            GPS_QUEUE,TRAFFIC_MANAGER,
                            LATITUDE and LONGITUDE>)
<Starting the web application on port 3000>
```

This is similar to the previous web application map used in *Chapter 13, Geolocalization Applications using GPS, NoSQL, and K3s Clusters,* but this one calls the GPS Queue service to get the current GPS coordinate that is running in the edge device and get data from the public endpoint of the **Traffic Manager** service that has to be accessed by using our custom proxy service to prevent CORS access restrictions. It also has the option to center the map at the beginning every time the page is loaded.

The web part uses the `map.html` file with the following code:

```
<!DOCTYPE html>
<html lang="en">
<head>
<Load Javascript libraries>
<Load page styles>
<body>
    <div id='map'></div>
<script>
    <Load Map in an initial GPS position>
    var marker
    var markers = []
    var osm = L.tileLayer(
    'https://{s}.tile.openstreetmap.org/{z}/{x}/{y}.png',
    {
        <Set Open Street Map Initial
        Configuration using Leaflet>
    });
    osm.addTo(map);
    setInterval(() => {
        $.getJSON("http://{{ GPS_QUEUE }}:3001/gps",
        function(gps) {
            <Delete current markers>
            <Get current position of your device
             and show it in the map>
                $.getJSON(
                    "http://{{ TRAFFIC_MANAGER }}:5000"+
                    "/traffic/unit/km/r/0.5/lat/<LATITUDE>"
                    "/lng/<LONGITUDE>", function(pos) {
                <This gets all the detected objects
                in a radius of 0.5 km>
                <For each object returned show it in
                the map using
                markPosition(object,lat,lng,o_type,warning)
                function>
```

```
                       });
             });
        }, 5000);


        <Configure the icons to visualize if an object is a
        person, car or a truck>
        function markPosition(object,lat,lng,o_type,warning)
        {
             <Create a maker with the appropriate Icon showing
             the object name, latitude, longitude, type of object
             and warning level>
        }
    </script>
    </body>
    </html>
```

This code basically centers the map with initial latitude and longitude coordinates, shows the current position of the device in a blue globe, and shows the detected objects with icons, showing the object name, the GPS coordinates, the type of object, and the warning level. It should look something like this:

Traffic Events

Leaflet | Map data © OpenStreetMap contributors, Imagery © Mapbox

Figure 14.3 – Driver current position

This shows the driver's current position in real time, while the vehicle is moving. The other possible visualization shows how detected objects appear across the map. This information is requested using the `proxy` service to visualize all detected objects by other drivers. This could represent a kind of **augmented reality (AR)**, something similar to what Waze does with its application. The visualization looks like this:

Traffic Events

Figure 14.4 – Detected object's current position and warning message

If you click inside the detected object, it will show the current GPS coordinate, the type of object, and a warning message. There are several objects included in this default implementation. The implementation includes car, truck, and person detection as possible obstacles and potential warnings for a driver. You can see the following icons on the map:

Figure 14.5 – Car, truck, and person icons shown in Traffic Map

By default, our web application updates the objects every 5 seconds within a radius of 0.5 kilometers. Those values can be customized to satisfy your own solution. Now, let's deploy our Traffic Map web application by executing the next commands:

1. Create a `traffic-map` deployment by running the following command:

```
$ cat <<EOF | kubectl apply -f -
apiVersion: apps/v1
kind: Deployment
metadata:
  creationTimestamp: null
  labels:
    app: traffic-map
  name: traffic-map
spec:
  replicas: 1
  selector:
    matchLabels:
      app: traffic-map
  strategy: {}
  template:
    metadata:
      creationTimestamp: null
      labels:
        app: traffic-map
    spec:
      containers:
      - image: sergioarmgpl/traffic_map
        name: traffic-map
        imagePullPolicy: Always
        env:
        - name: LATITUDE
          value: "<YOUR_LATITUDE_COORDINATE>"
        - name: LONGITUDE
          value: "<YOUR_LONGITUDE_COORDINATE>"
        - name: GPS_QUEUE
          value: "localhost" #<GPS_QUEUE_IP>
```

```
          - name: TRAFFIC_MANAGER
            value: "<TRAFFIC_MANAGER_IP>"
          resources: {}
     status: {}
     EOF
```

This deployment has the following environment variables:

- `LATITUDE`: Initial GPS latitude coordinate to center your map.

- `LONGITUDE`: Initial GPS longitude coordinate to center your map.

- `GPS_QUEUE`: IP address endpoint of the `gps-queue` service. In this case, because this runs locally, it is set by default as `localhost`.

- `TRAFFIC_MANAGER`: IP address endpoint of your **Traffic Manager** application. In this case, because of the use of the `proxy` service, we can call it using `localhost`, which prevents the CORS restriction.

Important Note

To check the code and create your own container of `traffic_map`, you can check the next link:

`https://github.com/sergioarmgpl/containers/tree/main/traffic-map/src`

We have now deployed the **Traffic Map** web application on our edge device. Let's move on to run our object detection system at the edge to perform our computer vision, in the next section.

Detecting objects with computer vision using OpenCV, TensorFlow Lite, and scikit-learn

The service that performs computer vision is contained in the `detect.py` file. This will run on our edge device. Let's explore the code inside this file before preparing our device to run this program, as follows:

```
<Imported libraries to run OpenCV in TensorFlow Lite>

#Array to map detected objects
obj_values = {"car":1,"cat":2,"person":3
,"dog":4,"semaphore":5,"truck":6,"other":1000}

def run():
```

```
    <Initialize Video Capture for the camera>
    <Set screen size to capture>
    <Initialize the object detection model>
    #Array to store detected objects
    items = []
    while Camera is Opened:
      detection_result = detector.detect(input_tensor)
      items.clear()
      <store detected objects in the items arrays>
      <Show the FPS evaluated>
      <Count objects detected per type of object>
      <Get the classification of each object calling
      /predict endpoint from the gps-api>
      if the warning count of the group <= 2:
          <A real warning is detected
          we push this information calling
          /traffic/event and warning is incremented>
      if warning:
          <show unique objects found
          warning is set to zero>
      else:
          <show No warnings>

      if <ESC key is pressed>:
        <break the cycle>
      <Set cv2 window size to show the capture>
    <Close the Camera Capture>
    <Destroy all windows>

def main():
  <Parse parameters to run the program>
  <Call run() function to start analyzing video capture>

if __name__ == '__main__':
  <call the main() function of the program>
```

This code starts the video capture and then sends this image in a format that TensorFlow Lite can analyze. TensorFlow Lite detects coordinates where objects are detected and classifies the objects with a label that is their name. This program will use the `efficientdet_lite0_edgetpu_metadata.tflite` model. In this case, we are focusing on the car, person, dog, semaphore, and truck objects. These objects represent obstacles for drivers and represent a level of warning. If the detected object is different than these objects, it's classified as `other` and it's omitted as a warning. If you want to add more objects to the list, you just have to modify the `obj_values` array with new values, as in the following example:

```
obj_values = {"car":1,..,"other_object":7,..."other":1000}
```

In each loop of this program, the detected objects are counted by groups and stored in the `items` array. Then, if one of these groups detects more than one object and the group is one of the identified objects in the `obj_values` array, the detected objects in the group are counted as potential object obstacles that represent warnings for drivers. To calculate the warning level, the script calls the `inference` API, and then, if a warning is detected, it calls the `traffic-map` service using the `proxy` service previously installed using the `http://localhost:5000/traffic/event` URL. Every time the proxy is called, the requests will be sent to the public endpoint of the `traffic-manager` service deployed in the cloud. Then, after the object analysis, the `items` array is cleared and the output summarizing the detected objects is shown in a blue box using OpenCV. It will look like this:

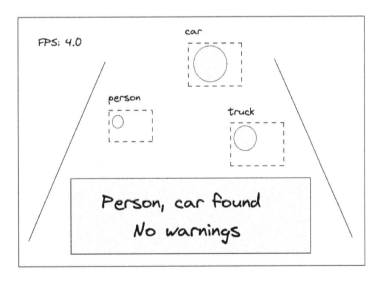

Figure 14.6 – Object detection screen

This output also shows the detected objects marked with a red rectangle with the name of the detected object. In the upper-left corner, you will see the number of FPS analyzed. Our warning box will show two types of messages: either the group of objects found (for example, **person, car found**) or that there are no detected objects—this will show the message **No warnings**. The service closes if you press the *Esc* key. To install the object detection service in your edge device, execute the following steps:

1. Connect your edge device to a network that you can access.

2. Log in to your edge device, like so:

    ```
    $ ssh your_user@<EDGE_DEVICE_IP>
    ```

    You can get the IP address of your device by running the following command:

    ```
    $ ifconfig wlan0
    ```

    You can run it by connecting your device to an HDMI screen and connecting a keyboard and mouse to your device.

3. Clone the repository by running the following code:

    ```
    $ git clone https://github.com/PacktPublishing/Edge-
    Computing-Systems-with-Kubernetes
    $ cd Edge-Computing-Systems-with-Kubernetes/ch14/code/
    python/object_detection
    ```

4. Install missing dependencies to run OpenCV and the camera, like so:

    ```
    $ /bin/bash install_deps.sh
    ```

5. Configure the device to run the object detection program, as follows:

    ```
    $ /bin/bash setup.sh
    ```

6. Run the script to install desktop shortcuts, like so:

    ```
    $ /bin/bash install_shortcuts.sh
    ```

> **Important Note**
> Take a look at the files with a .desktop extension that call the run.sh script and the files with a .desktop extension that start the detection application and the local web Traffic Map application. These files are located in the ch14/code/python/object_detection directory.

7. Test the installation by clicking on the new **Detector** desktop shortcut.

8. Test the local Traffic Map application by clicking on the **Traffic** desktop shortcut. This will open Chromium at `http://localhost:5000`.

9. Reconfigure your wireless network to use the access point connection of your smartphone and reset your `/etc/wpa_supplicant/wpa_supplicant.conf` configuration file by removing the `network {}` entries to use your smartphone internet connection.

> **Important Note**
> For more information, you can check the next link:
> `https://wiki.archlinux.org/title/wpa_supplicant`

10. Now, you can configure your touchscreen. In this case, we are using the Miuzei LCD 4.0-inch HDMI display, which flips the screen. For this, execute the following commands:

```
$ sudo rm -rf LCD-show
$ git clone https://github.com/goodtft/LCD-show.git
$ chmod -R 755 LCD-show
$ cd LCD-show
$ sudo ./MPI4008-show
```

11. Now, restart your device by running the following command:

```
$ sudo restart
```

12. Now, access the **Detect** shortcut to start the service to detect objects.

> **Important Note**
> You can accelerate the video-frame analysis by uncommenting the `--enableEdgeTPU` flag in the `ch14/code/python/object_detection/run.sh` file. Our detection code is based on the official Tensor Flow example that uses the Coral USB Accelerator device. This device is a TPU, which is a dedicated unit to process information using NNs. The configuration of the Coral device is out of the scope of this book. For more information, check the Coral USB Accelerator link in the *Further reading* section.

13. Start the Traffic Map application by clicking on the **Traffic** shortcut. If there are objects detected, they will appear 30 seconds later in the web application.

The last step is to deploy a public **Traffic Map** application to visualize all traffic in a radius area. For this, let's deploy the last service—**Traffic Map Public**—in the next section.

Deploying a global visualizer for the smart traffic system

The **Traffic Map Public** service is the static version of **Traffic Map** that only shows detected objects within a radius of 5 kilometers. This service is deployed in the cloud, so you should expect the same visualization as with the **Traffic Map** service, but the only missing part is that it doesn't show your real-time GPS position because it is static. The GPS position to take into consideration could be a GPS coordinate that is the center of the city that you want to monitor. In general, this web visualization could fit a static report for a municipality. The code is the same as for the **Traffic Map** web application, but the continuous update of the GPS position is omitted. To deploy this service, run the following commands:

1. Create a `traffic-map` deployment by running the following command:

```
$ cat <<EOF | kubectl apply -f -
apiVersion: apps/v1
kind: Deployment
metadata:
  creationTimestamp: null
  labels:
    app: traffic-map-public
  name: traffic-map-public
spec:
  replicas: 1
  selector:
    matchLabels:
      app: traffic-map-public
  strategy: {}
  template:
    metadata:
      creationTimestamp: null
      labels:
        app: traffic-map-public
    spec:
      containers:
      - image: sergioarmgpl/traffic_map_public
        name: traffic-map-public
        imagePullPolicy: Always
        env:
        - name: LATITUDE
```

```
            value: "<YOUR_LATITUDE_COORDINATE>"
          - name: LONGITUDE
            value: "<YOUR_LONGITUDE_COORDINATE>"
          - name: TRAFFIC_MANAGER
            value: "<TRAFFIC_MANAGER_IP>"
          resources: {}
  status: {}
  EOF
```

This deployment has the following environment variables:

- LATITUDE: Initial GPS latitude coordinate to center your map.

- LONGITUDE: Initial GPS longitude coordinate to center your map.

- GPS_QUEUE: IP address endpoint of the gps-queue service. In this case, because this runs locally, it is set by default as localhost.

- TRAFFIC_MANAGER: IP address endpoint of your **Traffic Manager** service. In this case, because of the use of the proxy, we can call it using localhost, which prevents the CORS restriction.

> **Important Note**
> To check the code and create your own container of traffic-map-public, you can check the next link:
> https://github.com/sergioarmgpl/containers/tree/main/traffic-map-public/src

2. Now, let's create a service for this deployment as a LoadBalancer. This IP address will be the endpoint to access the Traffic Map public web application. The code is illustrated in the following snippet:

```
$ cat <<EOF | kubectl apply -f -
apiVersion: v1
kind: Service
metadata:
  creationTimestamp: null
  labels:
    app: traffic-map-public
  name: traffic-map-public-lb
spec:
```

```
        ports:
        - port: 3000
          protocol: TCP
          targetPort: 3000
        selector:
          app: traffic-map-public
        type: LoadBalancer
    status:
      loadBalancer: {}
    EOF
```

> **Important Note**
>
> To troubleshoot your deployments, you can use the $ kubectl logs pod/<POD> -f
> <CONTAINER_NAME> command. This will show you some useful outputs to troubleshoot services.

3. Get the load balancer IP for your traffic-map-public deployment with the following command:

    ```
    $ TRAFFIC_MAP_PUBLIC="$(kubectl get svc traffic-map-
    public -o=jsonpath='{.status.loadBalancer.ingress[0].
    ip}')"
    ```

    You can see the value of the TRAFFIC_MAP_PUBLIC environment variable by running the following command:

    ```
    $ echo $TRAFFIC_MAP_PUBLIC
    ```

    Note that it takes some time after the IP address of the load balancer is provisioned. You can check the state of the services by running the following command:

    ```
    $ kubectl get svc traffic-map-public-lb
    ```

    Wait until the EXTERNAL_IP environment variable is provisioned.

4. Access the Traffic Map public application at http://<TRAFFIC_MAP_PUBLIC>:3000.

Now everything is running, try to fill the system with data and drive your car with your edge device to capture objects. You will then see the objects in the system in a few seconds. Take a look at the *Further reading* section, where there are a lot of materials that you can explore to create your system. But now, it's time to summarize what we learned. Let's move on to the *Summary* section.

Summary

In this chapter, we learned how you can use AI to analyze video captured by cameras, to detect objects that potentially represent obstacles for drivers. This was implemented to run at the edge on a Raspberry Pi, using the power of Kubernetes with K3s. With this approach, we created a decoupled system that could be easier to upgrade using containers. We also learned how this kind of system can be used in real-world scenarios to monitor traffic behavior to improve driver safety. Across this implementation, we also learned how this kind of system is distributed across the edge and the cloud to process and show information locally to drivers to improve their driving experience. In the last chapter, we are going to give an easy method to organize and design fast your own edge computing system using a diagram called the edge computing design system canvas.

Questions

Here are a few questions to validate your new knowledge:

- How are AI, ML, and computer vision related to each other to design smart traffic systems?
- How do TensorFlow Lite and scikit-learn work to detect objects and perform predictions?
- How does computer vision work running at the edge?
- How can you distribute data across the edge and the cloud?
- How can you use Python to build a computer vision system?
- How can you use K3s to design distributed systems that detect objects in real time?

Further reading

You can refer to the following references for more information on the topics covered in this chapter:

- *What is artificial intelligence (AI)?*: https://www.techtarget.com/searchenterpriseai/definition/AI-Artificial-Intelligence
- *Agents in Artificial Intelligence*: https://www.geeksforgeeks.org/agents-artificial-intelligence and https://www.educba.com/agents-in-artificial-intelligence
- *Smart Traffic Management: Optimizing Your City's Infrastructure Spend*: https://www.digi.com/blog/post/smart-traffic-management-optimizing-spend
- *Markers with Custom Icons*: https://leafletjs.com/examples/custom-icons
- *MLOps Using Argo and K3s*: https://github.com/sergioarmgpl/mlops-argo-k3s

- *YOLO and Tiny-YOLO object detection on the Raspberry Pi and Movidius NCS*: `https://pyimagesearch.com/2020/01/27/yolo-and-tiny-yolo-object-detection-on-the-raspberry-pi-and-movidius-ncs`

- *TensorFlow Lite example apps*: `https://www.tensorflow.org/lite/examples`

- *TensorFlow Hub*: `https://tfhub.dev`

- Get models for TensorFlow Lite: `https://www.tensorflow.org/lite/models`

- *Edge Analytics in Transportation and Logistics Space: A Case Study*: `https://www.skillsire.com/read-blog/174_edge-analytics-in-transportation-and-logistics-space-a-case-study.html`

- *Tutorial to set up TensorFlow Object Detection API on the Raspberry Pi*: `https://github.com/EdjeElectronics/TensorFlow-Object-Detection-on-the-Raspberry-Pi`

- *TensorFlow-Lite-Object-Detection-on-Android-and-Raspberry-Pi*: `https://github.com/EdjeElectronics/TensorFlow-Lite-Object-Detection-on-Android-and-Raspberry-Pi`

- *TensorFlow Lite Python object detection example with Raspberry Pi*: `https://github.com/tensorflow/examples/tree/master/lite/examples/object_detection/raspberry_pi`

- *Python Project – Real-time Human Detection & Counting*: `https://data-flair.training/blogs/python-project-real-time-human-detection-counting`

- Coral USB Accelerator: `https://coral.ai/products/accelerator`

- Edge TPU simple camera examples: `https://github.com/google-coral/examples-camera`

- *Use NGINX as a Reverse Proxy*: `https://www.linode.com/docs/guides/use-nginx-reverse-proxy`

- *Movidius on Mac OS*: `https://github.com/acharroux/Movidius-On-MacOS`

- *NCS-Pi-Stream*: `https://github.com/HanYangZhao/NCS-Pi-Stream`

- *Intel® Neural Compute Stick 2 (Intel® NCS2)*: `https://www.intel.com/content/www/us/en/developer/tools/neural-compute-stick/overview.html`

- *Deep Surveillance with Deep Learning – Intelligent Video Surveillance Project*: `https://data-flair.training/blogs/deep-surveillance-with-deep-learning-intelligent-video-surveillance-project`

- *Road Lane line detection – Computer Vision Project in Python*: `https://data-flair.training/blogs/road-lane-line-detection`

- *Raspberry Pi and Movidius NCS Face Recognition*: `https://pyimagesearch.com/2020/01/06/raspberry-pi-and-movidius-ncs-face-recognition`

- *OpenVINO, OpenCV, and Movidius NCS on the Raspberry Pi*: `https://pyimagesearch.com/2019/04/08/openvino-opencv-and-movidius-ncs-on-the-raspberry-pi`

- *Speed up predictions on low-power devices using Neural Compute Stick and OpenVINO*: `https://towardsdatascience.com/speed-up-predictions-on-low-power-devices-using-neural-compute-stick-and-openvino-98f3ae9dcf41`

- *Deep Learning with Movidius NCS (pt.4) Installing NCSDK on a Rock64*: `https://www.youtube.com/watch?v=AXzIYk7-lr8`

- *Glyph-based video visualization on Google Map for surveillance in smart cities*: `https://jivp-eurasipjournals.springeropen.com/articles/10.1186/s13640-017-0175-4`

- *Looking-In and Looking-Out of a Vehicle: Computer-Vision-Based Enhanced Vehicle Safety*: `https://escholarship.org/content/qt2g6313r2/qt2g6313r2_noSplash_81ae2290f201a6b25e8eecc8a1142845.pdf?t=lnpgaj`

- *Install Touch Screen and Touch Calibration Program for Raspberry Pi*: `https://www.gechic.com/en/raspberry-pi-install-touch-monitor-and-touch-calibrator-driver`

- *Rotating a Raspberry Pi 4 Touch Monitor*: `https://www.interelectronix.com/rotating-raspberry-pi-4-touch-monitor.html`

- *Calibrating Touchscreen*: `https://wiki.archlinux.org/title/Calibrating_Touchscreen`

15
Designing Your Own Edge Computing System

Sometimes, the success of a project is not the technology – it is the way that you design and execute it. Edge computing systems can start as a small startup idea, so you can use it to reference the lean canvas business plan template to do the first draft of the idea that you have to create the system. But what if we have some similar template adapted to edge computing? This is where the Edge Computing System Design Canvas can help you. The idea of this diagram is to give you a tool to create the first draft of all the things you need to create an edge computing system, and you can consider the chapters of this book as building blocks to create your own. In this chapter, we are going to explore cloud providers that you can use to host your services, some best practices to take into consideration, software that you can explore to build your edge computing system, and other use cases that you can explore to create a system if it's not covered in this book.

In this chapter, we're going to cover the following main topics:

- Using the edge computing system design canvas
- Using managed services from cloud providers
- Existing hardware for your projects
- Exploring complementary software for your system
- Recommendations to build your edge computing system
- Exploring additional edge computing use cases

Technical requirements

For more details, check out this resource on GitHub: `https://github.com/PacktPublishing/Edge-Computing-Systems-with-Kubernetes/tree/main/ch15`

Using the edge computing system design canvas

The edge computing system design canvas is based on the lean canvas business plan template, with the idea to have a tool to help people create and organize their edge computing systems by filling in a sheet of paper with what they need to start building their systems. Let's take a look at the different areas that our canvas template covers:

Edge Computing System Design Canvas

1. Purpose	2. Features	9. Edge	12. Cloud
3. Challenges			
	6. Automation	10. Devices	13. Communication
4. People			
	7. Data		
5. Costs	8. Security	11. Sensors	14. Metrics

Figure 15.1 – Edge computing system design canvas

Our template covers 14 areas that you can consider while designing the systems. First, you have to define the purpose of the system and the features to implement. Then, while filling this template, you can annotate in parallel the challenges, people, and costs to build the system. After that, you can define whether you are going to use automation in your system. In this category, we can talk about CI/CD pipelines and versioning. Later, you have to define how to manage your data, in which format, and then what security practices you are going to implement. The last two sections include what you are going to run at the edge and what type of devices and sensors you will use. Finally, you have to define which parts of the system are going to run in the cloud, how to communicate between the edge and the cloud, the metrics to collect from the edge, and which metrics are going to persist at the edge or in the cloud.

Now, let's explore some questions that you can use to fill in this sheet of paper. The idea is that you must fill it in no more than 10 minutes, similar to how the lean canvas works.

Purpose

As you know, you have to define why you want to build this system – that is, the main reason for this system to exist. You can discuss the following questions with your team:

- What is the purpose of the system?

- What is it going to do?

- What problem will it resolve?

Features

Here, you can list the top five features of your system. You can describe them in terms of the functionality of system attributes such as availability, reliability, and so on. Some questions that you can ask are as follows:

- What are the main features of your system?

- What functionalities do these features bring to the clients?

- What are the unique features of this system?

Challenges

Here, you have to detect the happy paths and potential blockers to build your system. Some questions that you can ask are as follows:

- What are the challenges of running software at the edge or in the cloud?

- What challenges will be faced when edge devices communicate with the cloud in the system?

You can complement these questions with other ones that evaluate the level of complexity of other technical areas to build the system.

People

Here, you have to evaluate people that are working in the system and define basic skills for future hiring. You can ask questions such as the following:

- What are the necessary skills to build the systems?

- How many people are necessary to build the system?

- How will the project be managed? Will this be in quarters, semesters, 2-week sprints, or in another way?

Costs

Here, you have to calculate possible costs to buy hardware, run third-party services, and more. You can ask questions such as the following:

- What is the cost of my devices?
- What is the cost of my sensors?
- What is the cost of my cloud provider?
- Who are my hardware providers?
- What additional costs do I have to consider?

Automation

Here, you have to evaluate the automation processes and code versioning. This is where you can fill in all the things related to CI/CD pipelines, data pipelines, GitOps, testing, and more. You can ask questions such as the following:

- What process is going to be automated?
- How will code versioning be implemented?
- Do you need CI/CD or GitOps?
- How will software testing be implemented?

Data

Here, you have to define how to manage data. This includes the format, databases, data ingestion, storage, and more. You can ask the following questions to define how to manage data in your system:

- Does the system use NoSQL databases?
- Does the system use SQL databases?
- What type of data (JSON, CSV, and so on) the system is going to use?
- What characteristics does my database need? This includes high availability, persistence, concurrency, partition tolerance, and others. You can use the CAP theorem to choose the best database to fit your needs.

Security

Here, you can evaluate the security of data and the services. This book doesn't cover this topic in particular, but you can ask the following questions to evaluate some minimal aspects of security within your system:

- Which security strategies are going to be implemented in your system?
- Where does data encryption need to be used in the system?
- How does system authentication work in the system?

Edge

In this section, you have to list and decide which devices are going to run at the edge. Here, you can find ARM devices and edge clusters. You have to decide which technologies you are going to use to run on your edge devices. You can ask the following questions to evaluate this:

- What is going to run at the edge?
- Which software is going to run on your devices?
- Does the system need a single or multi-node cluster running at the edge?
- Does the software run using virtual machines, containers, binaries, or something else?

Devices

This section is related to listing the possible devices to use in your systems and the additional hardware that you can use with them. You can ask the following questions to gather an initial list of possible devices that you can use:

- What type of processor will your devices use? ARM or x86_64?
- What additional hardware does my device need to use?
- How will the devices be powered? Using batteries or DC?
- How will the devices manage local time?
- What amount of memory for your firmware and data storage will be available for your device?

Sensors

The goal of this section is to list possible sensors and how to get data from them. Then, you must transform this information into metrics or variables to measure the environment. You can use the following questions to analyze the things related to sensors:

- Which sensors are you going to use?

- What are the sensors going to measure?

- Do the sensors need a source of power? What type of power do they need?

- How will the sensors be calibrated?

Cloud

This section is designed to evaluate which parts of the system have to run in the cloud, what managed services you are going to use if necessary, and if there are third-party services that could be used to reduce and simplify the time implementation of your system. To evaluate this, you can ask the following questions:

- What cloud provider fits your system needs the most?

- What managed services does the system need?

- Are there any third-party services that could be critical to use in the system?

Communication

This section is the result of communicating with the edge and the cloud layer. This is where you will define how the layers will communicate with each other, which protocols you are going to use, whether the communication is in real time or not, and whether your devices will use special protocols to communicate with each other. To fill in this section, you can ask the following questions:

- How will the edge devices transfer data to the cloud?

- What type of communication is going to be used to communicate with the edge devices and the cloud? This could be sockets, the REST API, gRPC, and so on.

- Does the system use Lora, Wi-Fi, Bluetooth, Sigfox, or other protocols to communicate to your devices at the edge or on the cloud?

- Will the communication be synchronous or asynchronous to store data?

Metrics

Your sensors at the edge generate data that will be transformed into metrics to be shown in a dashboard. However, these metrics to be defined. The goal of this section is to define the metrics to use and visualize them. These metrics are created using the edge recollected data. To define these metrics, you can ask the following questions:

- What type of metrics is the system going to collect? Golden metrics, weather metrics, or something else?

- Which metrics will be generated and used in the system? This can include latency, temperature, speed, and so on.

- How is the system going to visualize the collected data?

- Is the system going to use dashboard software to visualize data, such as Grafana or similar?

Please use the different chapters of this book as building blocks to create your system. You can use the templates at `https://github.com/PacktPublishing/Edge-Computing-Systems-with-Kubernetes/blob/main/ch15/docs/EdgeComputingSystemDesignCanvas.pdf` that are ready to print and design your edge computing system. Now, it's time to look at the relevant managed services from the top three cloud providers.

Using managed services from cloud providers

It is important to choose the right cloud provider. Several cloud providers are available, but the top three are **Amazon Web Services** (**AWS**), Google Cloud, and Azure. Let's look at the different managed services that you can use with these cloud providers:

- **AWS**: You can use virtual machines with EC2 and Graviton 2 ARM instances to test software that you will run at the edge. Fargate is a service that you can use to deploy applications in containers. It provides several options. For instance, you can scale the service automatically, something similar to what Kubernetes does. **Elastic Kubernetes Service** (**EKS**) is the managed service of AWS for Kubernetes. It's a very strong solution for EKS, but compared to other services, you have to do more manual steps for certain tasks, such as scaling the solution. Talking about databases, you can use Aurora as a MySQL or Postgres instance. You can use other managed services based on Redis or Elastic Cache. For file storage, you can use S3 services. Finally, for complete serverless solutions, you can deploy Lambda functions, which run small portions of code on demand. AWS also has an IoT platform to connect devices running at the edge. AWS offers some certified devices to work with its platform. The official website is `https://aws.amazon.com`.

- **Google Cloud**: This cloud provider includes virtual machines, which are the same as AWS EC2 instances. This service is part of the Compute Engine services called VM instances. Google Cloud also offers ARM instances via the Tau instance type. It provides the Cloud Run service, which runs containers, and **Google Kubernetes Engine** (**GKE**), which is a Kubernetes-managed service that is simpler to manage than EKS, and it's a much more stable solution. For databases, you can use Memory Store, a self-managed Redis service, and Cloud SQL, which is similar to AWS Aurora. However, in terms of databases, Google has less prebuild options than AWS, though it works pretty similarly. It can run as MySQL, Postgres, and SQL server. It has its own way to manage storage using cloud storage and works similar to S3 by using buckets to store information. Finally, it also has a serverless capability with Cloud Functions, which are similar to AWS Lambda. One of the main differences in Google Cloud is that its service definitions are compatible with open source projects. For example, Flask is compatible with Cloud Functions, and Cloud Run is compatible in some way with Knative. One of the major advantages of using Google Cloud is its compatibility with open source projects. Google Cloud also has its own IoT solution, similar to AWS, but it also works with some open source hardware and devices, such as Coral USB Accelerators or the Coral Dev Board. The official website can be found at `https://cloud.google.com`.

- **Azure**: This feels like a combination of AWS and Google Cloud and provides similar tools. It has virtual machines services, and it also supports ARM processors with the Dpsv5 and Epsv5 instances. It also has **Azure Kubernetes Service** (**AKS**), which is the managed Kubernetes service for Azure. AKS has some disabled features that are a little bit complicated to configure, even with the correct configuration, so it feels less mature compared to AWS and Google Cloud. It is also a little bit more expensive, but it depends on the quantity and size of cluster nodes that your system needs. AKS is less mature than AWS and Google Kubernetes managed services. Azure also has Azure Container Instances, which are used to run containers such as AWS Fargate and Google Cloud Run. For databases, it offers Azure Cosmos DB, which provides a database-managed service such as AWS Aurora or Google Cloud SQL. This database offers compatibility with Cassandra, SQL Server, MongoDB, and Gremlin, which is similar to Neo4J. Cosmos DB is more like the NoSQL version of Aurora and Cloud SQL. It also has an enterprise Redis service by default. Talking about serverless functions, it provides Azure Functions, which support languages such as Python and TypeScript, and some proprietary languages owned by Microsoft such as C# and PowerShell scripting. Azure in the context of IoT has more options to connect your devices and has a lot of certified hardware designed to run with Azure. It feels like this platform is frequently innovating. The official website can be found at `https://azure.microsoft.com`.

For this specific kind of book, you can also consider the **Civo** cloud, which provides a managed K3s service that you can use to play around with K3s. The official website can be found at `https://www.civo.com`.

> **Important Note**
> Take a look at the official website of each cloud provider for updates about their current services.

All this information was just a brief introduction to what these cloud providers offer, so not all the facilities of each cloud provider have been covered. Maybe you are thinking about who the best cloud provider is. The answer depends on what you prefer for certain solutions because of current service contracts, previous software adoption, and so on. Some cloud provider services are better in some situations than others, and your team will have to spend some time evaluating this. The cost of a service can change depending on the size of the services on each cloud provider. To help with your decision, you can think about the following questions:

- Is the managed services price of the provider fair to substitute for a self-managed service that the system is planning to use?
- Is the learning curve of the managed service adoption worthwhile and will it affect the deadline of the project?
- Does the cloud provider include the majority of services that need to be implemented in the system without using another cloud provider?
- Does the cloud provider include support and good documentation to use their services?
- Does the adoption of the selected cloud provider allow you to keep running your applications without you having to make many modifications to the source code of your application?

These are some questions that you can ask the team of your project, and they could act as a good starting point to evaluate a cloud provider. Now, let's explore some hardware that you can use in your edge computing systems.

Existing hardware for your projects

There is plenty of hardware that you can use for your edge computing projects. Let's look at a small list of hardware that you can use for your projects. The following list includes microcomputers such as the Raspberry Pi and microcontrollers such as Arduino:

- **Coral Dev Board**: This is a board designed by Google that uses the Coral Accelerator to run ML applications. It is a reasonable size and provides processing power to run machine learning applications. For more information, check out `https://coral.ai/products/dev-board`.
- **Rock Pi**: This device is similar to a Raspberry Pi but includes a Mali GPU, which can be used to process machine learning applications. It also has other board versions that you can use to run at the edge. For more information, check out `https://rockpi.org`.

- **Pine64**: This is a community platform that creates boards that have ARM processors. It also has another product that can be used at the edge, similar to the Raspberry Pi. For more information, check out `https://www.pine64.org`.

- **ESP32**: This is a commonly used microcontroller that you can program to read information with sensors at the edge. There are plenty of distributors with a lot of variations of the ESP32 that already integrate sensors. For more information, check out `https://heltec.org/proudct_center/esp-arduino`.

- **MicroPython**: This board is designed to run Python. It has a lot of features that can be used to quickly prototype a device to capture data at the edge. For more information, check out `https://micropython.org`.

- **NVIDIA Jetson Nano**: This device is designed by NVIDIA and has a powerful GPU. It has a lot of power to run processes and it could be a good option for running intensive tasks, including machine learning. For more information, check out `https://developer.nvidia.com/embedded/jetson-nano-developer-kit`.

Note that there are devices that could just be used to prototype a solution, though it is not recommended to run them in a production scenario. Check out the *Further reading* section to find other devices. Now, let's explore some complementary software that you can use at the edge to create your system.

Exploring complementary software for your system

There are other pieces of software that you can use if some of the examples in this book don't fit your system needs. Some examples are as follows:

- **Crossplane**: This is used to deploy infrastructure using Kubernetes. Crossplane can give you the abstraction to do this. For more information, check out `https://crossplane.io`.

- **Thanos**: This is a Prometheus cluster that you can use to scale your Prometheus services. For more information, check out `https://thanos.io`.

- **Argo**: This is a whole ecosystem that you can use to implement GitOps, workflows, and event management. It is a powerful piece of software. Argo can also run on ARM devices. For more information, check out `https://argoproj.github.io`.

- **Containerd**: If K3s is too big for your solution, then you may wish to use containers. Containerd can give you this abstraction without extra services. For more information, check out `https://containerd.io`.

- **Rancher**: This is a Kubernetes distribution that you can use to manage all your clusters at the edge so that you can have a single dashboard application to manage and monitor all your clusters in a single place. For more information, check out `https://rancher.com`.

- **KubeSphere**: This is similar to Rancher but has a different approach so that it's more developer-friendly; Racher is more operations-friendly. For more information, check out `https://kubesphere.io`.

- **OpenEBS**: This is an alternative to Longhorn that has pretty good support and options for storage. For more information, check out `https://openebs.io`.

- **KubeEdge**: This is a modification of K3s that's used to distribute your nodes across the cloud and the edge. It also supports MQTT protocols. For more information, check out `https://kubeedge.io`.

- **Akri**: This is a Kubernetes resource interface that can easily expose your devices in the tiny edge, such as cameras or USB devices, as resources in a Kubernetes cluster. For more information, check out `https://docs.akri.sh`.

You can also explore the graduated, incubated, and sandbox projects of CNCF at `https://www.cncf.io` and the landscape at `https://landscape.cncf.io` to explore more options that you can add to your project. Now, let's continue with some useful recommendations when creating your system.

Recommendations to build your edge computing system

Here is a list of recommendations that you can consider when designing your edge computing system:

- Take your time when designing the system. You can do this on paper, which will save you a lot of time when building your system.

- Measure the progress of building your system. Without measure, there is no pressure. You can use the Scrum and Kanban agile methodologies to manage the progress of your project. It's very important to plan.

- Invest time in making a **proof of concept** (**POC**) after deciding which technology, cloud provider, or a third party you will use. This will be critical to have a constant process when building your system.

- Invest time in documentation. This is the only way you don't lose knowledge if someone has left the job.

- Version the code of your projects. This is a healthy best practice to ensure you don't lose important code in your project.

- Use encryption. Evaluate the places where you can find sensible data in the system and encrypt it.

- Use secrets as a general rule. This book contains a lot of examples that don't use secrets to simplify the examples. However, in a real-world scenario, using secrets is a must.

- Think as a hacker. Sometimes, you have to think about the worst-case scenarios to consider how people can steal your information.

- Invest in professionals with experience but don't forget newbies. When hiring people for the project, pay attention to the experience that someone in the team can give you but remember that the youngest talents could have innovative ideas.

With that, you have a set of recommendations for building an edge computing system. Next, let's explore other use cases for edge computing.

Exploring additional edge computing use cases

To finalize this book, here are some use cases that you can explore and implement using edge computing technologies:

- **Healthcare**: In this system, data could be processed or analyzed locally. Sometimes, this information could be processed using artificial intelligence. This system can integrate local sensors and process information at the edge.

- **Industry 4.0**: This is related to the use of edge computing and IoT for manufacturing processes, where you can process information at the edge with ARM devices to reduce latency when interconnecting systems and data processing.

- **Autonomous vehicles**: This industry is constantly growing with the emerging market of electric cars. This use case employs cameras, augmented reality, and computer vision with the goal of cars driving autonomously.

- **Gaming**: This use case focuses on sharing the processing between the cloud and high-end user devices such as consoles to reduce the lag of video games.

- **Security**: In security use cases, cameras could be used for monitoring and detecting dangerous behaviors in people or to prevent robberies. These kinds of systems usually use object detection and artificial intelligence for this purpose.

- **Agriculture**: This could be applied to smart farms or gardens, to monitor plants and perform actions such as watering. This use case has some contact with IoT technologies and long-distance protocols such as LoRa.

- **Smart cities**: This use case has a lot of applications, such as smart traffic, which consists of monitoring traffic and its safety, thus improving the traffic flow in a city.

- **Logistics**: This use case improves the delivery time of packages, optimizes delivery routes and fuel consumption, and so on. This could present a competitive advantage for companies in the market.

There are plenty of other use cases that you can explore. Check out the *Further reading* section for more information. Now, let's summarize what we have learned in this chapter.

Summary

In this chapter, we learned about some complementary content for designing and implementing edge computing systems. First, we covered an edge computing system design canvas and asked some useful questions that could be used to quickly start designing your system. After that, we explored the top cloud providers with managed services that can be used with edge computing systems and some hardware that can be used for this purpose. Finally, we looked at some complementary software to use as recommendations and other use cases to explore. With all this information, you can organize and quickly start building an edge computing system. This content could be useful for organizing all the ideas behind your edge computing systems. Thank for reading this book – I hope you enjoyed it.

Questions

Here are a few questions to validate your new knowledge:

- How can you use the edge computing systems design canvas to design an edge computing system?
- What cloud providers can you use to complement your system?
- What complementary software or hardware you can use to build your system?
- What other use cases can be implemented with edge computing?

Further reading

Please refer to the following references for more information on the topics covered in this chapter:

- Cloud Native Computing Foundation: `https://www.cncf.io`
- Azure Certified Device catalog: `https://devicecatalog.azure.com`
- Azure IoT developer Kit: `https://microsoft.github.io/azure-iot-developer-kit`
- Adafruit: `https://www.adafruit.com`
- M5stack electronics store: `https://shop.m5stack.com/collections`
- EMQX, The Most Scalable MQTT Broker for IoT: `https://www.emqx.io`
- Seeed Studio IoT store: `https://www.seeedstudio.com`
- 12 Real-Life Edge Computing Use Cases: `https://www.scitechsociety.com/12-real-life-edge-computing-use-cases`
- Edge Analytics in Transportation and Logistics Space: A Case Study: `https://www.skillsire.com/read-blog/174_edge-analytics-in-transportation-and-logistics-space-a-case-study.html`

- How Kubernetes is shaping the future of cars: `https://thechief.io/c/editorial/how-kubernetes-is-shaping-the-future-of-cars`

- Edge Use Cases for Retail, Warehousing, and Logistics: `https://stlpartners.com/articles/edge-computing/edge-use-cases-for-retail-warehousing-and-logistics`

- Edge Computing Use Cases Driving Innovation: `https://www.section.io/blog/edge-compute-use-cases`

- IoT vs. Edge Computing: What's the difference: `https://developer.ibm.com/articles/iot-vs-edge-computing`

Index

C

O

P

Packt.com

Subscribe to our online digital library for full access to over 7,000 books and videos, as well as industry leading tools to help you plan your personal development and advance your career. For more information, please visit our website.

Why subscribe?

- Spend less time learning and more time coding with practical eBooks and Videos from over 4,000 industry professionals

- Improve your learning with Skill Plans built especially for you

- Get a free eBook or video every month

- Fully searchable for easy access to vital information

- Copy and paste, print, and bookmark content

Did you know that Packt offers eBook versions of every book published, with PDF and ePub files available? You can upgrade to the eBook version at packt.com and as a print book customer, you are entitled to a discount on the eBook copy. Get in touch with us at customercare@packtpub.com for more details.

At www.packt.com, you can also read a collection of free technical articles, sign up for a range of free newsletters, and receive exclusive discounts and offers on Packt books and eBooks.

Other Books You May Enjoy

If you enjoyed this book, you may be interested in these other books by Packt:

End-to-End Automation with Kubernetes and Crossplane

Arun Ramakani

ISBN: 9781801811545

- Understand the context of Kubernetes-based infrastructure automation
- Get to grips with Crossplane concepts with the help of practical examples
- Extend Crossplane to build a modern infrastructure automation platform
- Use the right configuration management tools in the Kubernetes environment
- Explore patterns to unify application and infrastructure automation
- Discover top engineering practices for infrastructure platform as a product

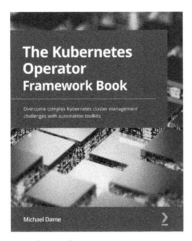

The Kubernetes Operator Framework Book

Michael Dame

ISBN: 9781803232850

- Gain insight into the Operator Framework and the benefits of operators
- Implement standard approaches for designing an operator
- Develop an operator in a stepwise manner using the Operator SDK
- Publish operators using distribution options such as OperatorHub.io
- Deploy operators using different Operator Lifecycle Manager options
- Discover how Kubernetes development standards relate to operators
- Apply knowledge learned from the case studies of real-world operators

Packt is searching for authors like you

If you're interested in becoming an author for Packt, please visit `authors.packtpub.com` and apply today. We have worked with thousands of developers and tech professionals, just like you, to help them share their insight with the global tech community. You can make a general application, apply for a specific hot topic that we are recruiting an author for, or submit your own idea.

Share Your Thoughts

Now you've finished *Edge Computing Systems with Kubernetes*, we'd love to hear your thoughts! Scan the QR code below to go straight to the Amazon review page for this book and share your feedback or leave a review on the site that you purchased it from.

https://packt.link/r/1-800-56859-2

Your review is important to us and the tech community and will help us make sure we're delivering excellent quality content.

www.ingramcontent.com/pod-product-compliance
Lightning Source LLC
LaVergne TN
LVHW081328050326
832903LV00024B/1066